T0223795

Pragmatic Circuits:
Signals and Filters

Pragmatic Circuits: Signals and Filters
William J. Eccles

ISBN (13 digits):978-3-031-79751-4   paperback
ISBN (13 digits): 978-3-031-79752-1   ebook

DOI: 10.1007/978-3-031-79752-1

A Publication in the Springer series
SYNTHESIS LECTURES ON DIGITAL CIRCUITS AND SYSTEMS #4

Series Editor: Mitchell Thornton, Southern Methodist University

Series ISSN: 1932-3166   print
Series ISSN: 1932-3174   electronic

10 9 8 7 6 5 4 3 2 1

Pragmatic Circuits:
Signals and Filters

William J. Eccles
Rose-Hulman Institute of Technology
Terre Haute, Indiana, USA

SYNTHESIS LECTURES ON DIGITAL CIRCUITS AND SYSTEMS #4

ABSTRACT

Pragmatic Circuits: Signals and Filters is built around the processing of signals. Topics include *spectra*, a short introduction to the *Fourier series*, design of *filters*, and the properties of the *Fourier transform*. The focus is on signals rather than power. But the treatment is still pragmatic. For example, the author accepts the work of Butterworth and uses his results to design filters in a fairly methodical fashion. This third of three volumes finishes with a look at spectra by showing how to get a spectrum even if a signal is not periodic. The Fourier transform provides a way of dealing with such non-periodic signals. The two other volumes in the ***Pragmatic Circuits*** series include titles on ***DC and Time Domain*** and ***Frequency Domain.***

These short lecture books will be of use to students at any level of electrical engineering and for practicing engineers, or scientists, in any field looking for a practical and applied introduction to circuits and signals. The author's "pragmatic" and applied style gives a unique and helpful "non-idealistic, practical, opinionated" introduction to circuits

KEYWORDS:

signals, spectrum, Fourier series, filter, Fourier transform, frequency domain

Contents

CHAPTER 9

Spectrum: More than just Rainbows

The rainbow is a spectrum of colors. It ranges from red to violet, as far as we can tell. Yet we know that there is "color" in the rainbow beyond that range, extending into the infrared and the ultraviolet.

If we understand light as having *wavelength* and hence its reciprocal *frequency*, we can say that there is a whole range of frequencies present in the colors of the rainbow. That range of frequencies is the *spectrum* of the rainbow.

When we talk about processing signals, we often talk about the *spectrum* of the signal. Sometimes that spectrum is very simple, a single frequency perhaps. Sometimes it is more complicated, the fundamentals and the overtones of a violin note, for example. And sometimes we really don't know the exact frequencies, as in speech. So in this last case we often resort to a descriptive range of frequencies that we call the spectrum.

How does this all apply to what we've been learning about circuits? That's what this chapter is about. We'll start with simple signals with just a few frequencies. Using these, we'll learn to plot a spectrum, which means we'll be back in the *frequency domain*. Then we'll take up more complicated spectra, including not just voltage but also power.

9.1 HARMONICALLY RELATED SOURCES

Most of the sources and signals that we have dealt with so far have been single-frequency signals. We've had DC sources such as

$$v_i(t) = 10\,\text{V},$$

and AC sources such as

$$v_i(t) = 5\cos(690t + 20°)\,\text{V}.$$

We've passed these through circuits like the generalized one in Fig. 9.1 to get outputs.

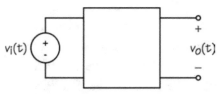

FIGURE 9.1: A system.

We've combined sources in a linear fashion:

$$v_i(t) = 10 + 5\cos(690t + 20°)\,\text{V}.$$

If our system is linear, we use superposition to find the output $v_o(t)$, finding the output due to each part of the input and then summing the results.

In this chapter we are more concerned with the sources themselves. Suppose our source is the sound of a trumpet playing an A (440 Hz). The trumpet certainly is not producing a pure sound described by $\cos(2\pi440t)$. Something is different about the sound. Likewise, a violin playing an A is not just a pure single frequency, and it certainly is different from a trumpet.

There's a certain *timbre* to the sounds of these instruments, something that makes a trumpet sound like a trumpet and a violin sound like a violin. Musicians call this difference *overtones*; engineers call it *harmonics*.

Suppose we have a source that contains the frequency 440 Hz and three harmonics of that frequency: 880 Hz, 1320 Hz, and 1760 Hz. The first of these (440) is called the *fundamental*; the rest are called *harmonics*. Harmonics are usually numbered, counting the fundamental as No. 1. So our signal consists of the fundamental (440 Hz), the second harmonic (880 Hz), the third harmonic, and so on. (Note that the fundamental is also the *first* harmonic.)

We usually find that harmonics diminish in amplitude as the frequency increases. So let's make our example waveform do this:

$$v_i(t) = 10\cos 2\pi440t$$
$$+6\cos(2\pi880t + 20°)$$
$$+2\cos(2\pi1320t + 50°)$$
$$+0.5\cos(2\pi1760t + 80°)\,\text{V}.$$

I've included a changing phase angle with each term.

What does this signal look like? Fig. 9.2 shows a plot of $v_i(t)$ over about two periods of the fundamental.

Now let's pass this signal through the filter shown in Fig 9.3. Recall that the gain of an op-amp circuit like this is 1 plus the ratio of the feedback impedance to the grounded

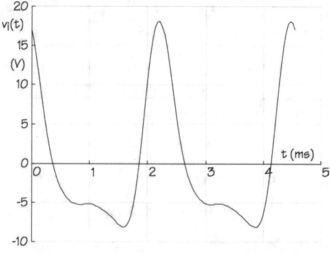

FIGURE 9.2: Input voltage.

impedance. As the frequency increases, the capacitor becomes more and more a short circuit. So the gain increases. Hence this filter passes high frequencies better. That makes it a *high-pass* filter.

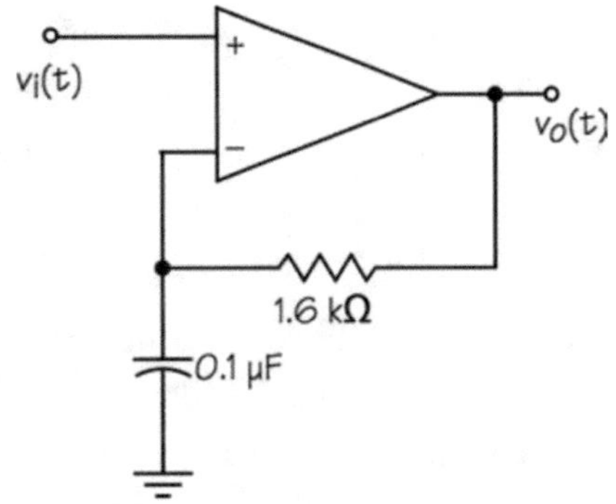

FIGURE 9.3: Filter circuit.

How do we find $v_o(t)$ for our circuit? If we want to work in the time domain, we take each of the four terms of $v_i(t)$ separately and find out what happens. In other words, we apply superposition, which we can do because our circuit is linear and the input is a linear combination of single-frequency signals.

But I suspect you would prefer to work in the phasor domain, which is a lot simpler when we are interested in the steady-state output. Here too we use superposition, but we must be careful. For example, the phasor for the fundamental is 10 $\angle 0°$ V, while the phasor for the second harmonic is 6 $\angle 20°$ V. It sure looks like we could combine them! Nope, never never never! These are phasors at two different frequencies and they must stay forever separate.

So in the phasor domain our analysis will use four different phasors:

$$V_{i1} = 10\angle 0° \text{ V at } 2\pi 440 \text{ rad/s},$$
$$V_{i2} = 6\angle 20° \text{ V at } 2\pi 880 \text{ rad/s},$$
$$V_{i3} = 2\angle 50° \text{ V at } 2\pi 1320 \text{ rad/s},$$
$$V_{i4} = 0.5\angle 80° \text{ V at } 2\pi 1760 \text{ rad/s}.$$

The gain of our filter is

$$gain = 1 + \frac{1600}{1/10^{-7}s} = 1 + \frac{1600}{1/j2\pi 10^{-7}f}$$
$$= 1 + j1.005 \times 10^{-3} f.$$

When we do all the math (I let Maple do the work), the phasor outcome is

$$V_{o1} = 10.93\angle 23.9° \text{ V at } 2\pi 440 \text{ rad/s},$$
$$V_{o2} = 8.01\angle 61.5° \text{ V at } 2\pi 880 \text{ rad/s},$$
$$V_{o3} = 3.32\angle 103.0° \text{ V at } 2\pi 1320 \text{ rad/s},$$
$$V_{o4} = 1.01\angle 140.5° \text{ V at } 2\pi 1760 \text{ rad/s}.$$

Putting these results back into the time domain and superimposing them gives us $v_o(t)$:

$$v_o(t) = 10.93\cos(2\pi 440t + 23.9°)$$
$$+ 8.01\cos(2\pi 880t + 61.5°)$$
$$+ 3.32\cos(2\pi 1320t + 103.0°)$$
$$+ 1.01\cos(2\pi 1760t + 140.5°)\text{ V}.$$

Note that the higher harmonics have been amplified more than the fundamental. What does the result look like? Fig. 9.4 is a plot of a couple of periods of both the output and the input signals.

Two things are worth noting. First, the harmonics have increased (the sharp peak is more pronounced and is pointier). Second, the whole signal is advanced slightly in time.

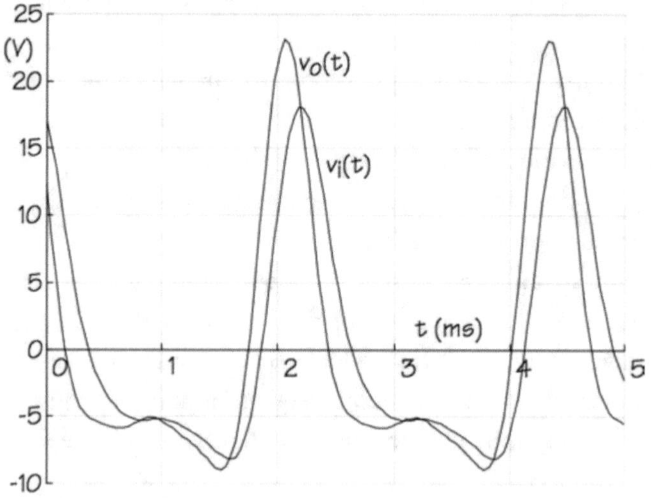

FIGURE 9.4: Input and output voltages.

9.2   "PLOTTING" SIGNALS

Consider a simple signal:

$$v(t) = A\cos(2\pi f_a t + \theta_a)\text{ V}.$$

This signal has four different things that could be variables:

A the amplitude of the cosine waveform;

f_a the frequency of the signal in hertz;

t the independent variable "time"; and

θ_a the phase angle, usually expressed in degrees, even though that's adding apples and oranges.

We generally think of t as the independent variable, one that we don't control. We plot the actual time function $v(t)$ against the independent variable t. So t seems to be somewhat sacred.

But is it? For example, when we are in the frequency domain, f is the independent variable and we express functions in exponential or phasor notations. So when it comes right down to it, there are four different variables. If we make a two-dimensional plot of this signal, there are a dozen different pairings that we could use for the ordinate and the abscissa of a plot.

Suppose we plot the amplitude A of our signal against the frequency f. Since both A and f are fixed, we get one point on the graph shown in Fig. 9.5.

How about phase versus frequency? Fig. 9.6 shows a plot of θ_a versus f.

Again, we get a single point. (It is common to draw a vertical line to join the single point to the axis.)

These drawings make better sense when we think phasors. The time-domain function is

$$v(t) = A\cos(2\pi f_a t + \theta_a)\,\text{V},$$

which transforms into the phasor

$$V(f) = A\angle\theta_a \text{ at } f_a$$
$$= Ae^{j\theta_a}\,\text{V}.$$

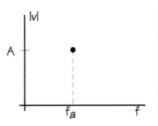

FIGURE 9.5: Amplitude.

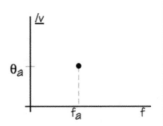

FIGURE 9.6: Phase.

So our plots are really of the characteristics of the phasor representation of the original signal. *Frequency* is the independent variable.

These drawings are a *spectrum* that represents the signal in the frequency domain. In simple terms, a spectrum is a graph of the magnitude or the phase angle of a signal plotted against frequency as the independent variable.

Consider the signal that we had in the previous section. It consisted of a fundamental at 440 Hz and three harmonics. The two-dimensional plots of amplitude and of phase versus frequency are shown in Fig. 9.7.

OK, now we know what a spectrum is, or at least what two examples look like. But as we shall see in the next section, these spectrums (spectra?) are not quite complete. There is more to the story. In fact, so far you've seen only half the plot.

9.3   SPECTRUM

We must think of a spectrum in terms of phasors. After all, a spectrum has frequency as the independent variable, and so does a phasor. We know that the phasor $Ae^{j\theta}$ represents in the frequency domain

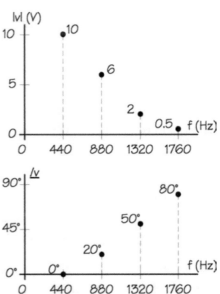

FIGURE 9.7: A spectrum, sort of.

the time function $A\cos(2\pi ft+\theta)$. To get from the frequency domain to the time domain, we go through the conversion as follows:

$$Ae^{j\theta_1} \ (A\angle\theta_1)$$
$$\Downarrow$$
$$Ae^{j(2\pi f_1 t+\theta_1)}$$
$$\Downarrow$$
$$\mathrm{Re}\left[Ae^{j(2\pi f_1 t+\theta_1)}\right]$$
$$\Downarrow$$
$$A\cos(2\pi f_1 t + \theta_1).$$

There is just one frequency appearing in this, namely, f_1. But the real-part operation is nonlinear. Half of the conversion from the complex exponential has been discarded.

9.3.1 Double-Sided Spectra

If we do the same job without using the real-part operation, we get something that's different. Let's use Euler's formula to convert the cosine into complex exponential form without the real-part operation:

$$\mathrm{A}\cos(2\pi f_1 t + \theta_1) = \frac{A}{2}e^{j(2\pi f_1 t+\theta_1)} + \frac{A}{2}e^{-j(2\pi f_1 t+\theta_1)}$$
$$= \frac{A}{2}e^{j\theta_1}e^{j2\pi f_1 t} + \frac{A}{2}e^{-j\theta_1}e^{-j2\pi f_1 t}.$$

What does this all say? Recall that we convert terms such as these into phasors by simply dropping the time term. Now look at the two terms:

- the original cosine is split into two parts;
- half the function is at the positive frequency f_1, while the other half is at the negative of that value;
- the two terms have the same phase angle, but one is positive and one is negative; and
- the amplitude of both terms is positive and half as large as that of the original cosine.

Let's plot our new result in Fig. 9.8. This spectrum is now double sided, including both positive and negative frequencies. The right-hand half is the familiar one for positive frequencies. The left-hand half is the *complex conjugate* of the right-hand half.

Do we need double stuff? (In Oreo® cookies, yes!) After all, you say, we really haven't added much. And you are about right for simple cases. But the simple case falls apart under some circumstances. Let's use an example, amplitude modulation, to show where the problem is.

Amplitude modulation (the popular AM band on radios) modifies (*modulates*) the amplitude of a radio-frequency carrier. The modulating signal is the information we wish to transmit. Mathematically, this looks like[1]

$$v(t) = (A \cos 2\pi f_{mod} t) \cos 2\pi f_{car} t.$$

The frequency f_{car} is the radio-frequency carrier. Its amplitude is a variable, shown in parentheses in the equation. This amplitude contains a modulating frequency f_{mod}. Let's use some numbers:

$$f_{car} = 1\,\text{MHz at } 0° \text{ and 10-V peak,}$$
$$f_{mod} = 1\,\text{kHz at } 0° \text{ and 2-V peak.}$$

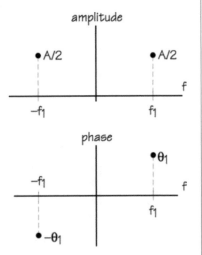

FIGURE 9.8: Discrete spectrum.

This gives us a signal in the time domain of

$$v(t) = (2 \cos 2\pi 10^3 t) 10 \cos 2\pi 10^6 t$$
$$= 20 \cos 2\pi 10^3 t \cos 2\pi 10^6 t \text{ V.}$$

How do we handle this in phasor notation? Can we write

$$V_{AM} = 20\angle 0° \angle 0° ?$$

No! No! 1000* No! Phasors of different frequencies can't be mixed like that! Hmmm, so now what? I guess we could try the real-part approach:

$$v(t) = 20 \text{Re}\left[e^{j2\pi 10^3 t} \right] \text{Re}\left[e^{j2\pi 10^6 t} \right] \text{V.}$$

But we get into mathematical trouble because the real-part operation is not commutative (i.e., we can't pull the arguments of both *Re* operations into a single *Re*:

$$\text{Re}[a + jb]\text{Re}[c + jd] = ac$$
$$\neq$$
$$\text{Re}[(a + jb)(c + jd)] = \text{Re}[ac - bd + jbc + jad]$$
$$= ac - bd.$$

[1]To truly be amplitude modulation, the modulating term should be (1 + Acos...). What is described here is really double-sideband suppressed-carrier modulation, a fairly rare form.

So that doesn't work. But Euler will rescue us and at the same time show that we have to use both the positive and negative components of frequency:

$$v(t) = 20 \left[\frac{e^{j2\pi 10^3 t} + e^{-j2\pi 10^3 t}}{2} \right] \left[\frac{e^{j2\pi 10^6 t} + e^{-j2\pi 10^6 t}}{2} \right]$$

$$= 5 \left[e^{j2\pi(10^3 + 10^6)t} + e^{j2\pi(10^3 - 10^6)t} \right.$$

$$\left. + e^{-j2\pi(10^3 - 10^6)t} + e^{-j2\pi(10^3 + 10^6)t} \right]$$

$$= 5 \left[e^{j2\pi(10^3 + 10^6)t} + e^{j2\pi(10^3 - 10^6)t} \right.$$

$$\left. + e^{j2\pi(10^6 - 10^3)t} + e^{j2\pi(10^6 + 10^3)t} \right] V.$$

(All phase angles are zero, so no "angle" terms appear in the result.)

There are *four* different frequencies present. A close examination of the terms shows

$$10^6 + 10^3 = 1.001 \ \text{MHz},$$

$$10^6 - 10^3 = 0.999 \ \text{MHz},$$

$$-10^6 + 10^3 = -0.999 \ \text{MHz},$$

$$-10^6 - 10^3 = -1.001 \ \text{MHz}.$$

These make a double-sided plot of the spectrum as shown in Fig. 9.9.

We clearly have a double-sided display. But when we tune our AM radio to a station at 1 MHz that is sending out a 1-kHz tone, is that negative stuff there? Does our AM radio really receive negative frequency? Well, frequencies in the real world are positive, but

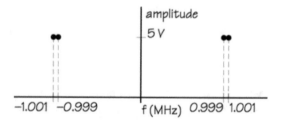

FIGURE 9.9: Modulation spectrum.

the way we do our mathematics requires that we use both sides. So is the negative stuff really there? I'll leave that to the philosophers.

Aha! you say, I can remember some high-school trig and I know there's a product-of-cosines formula. Why not use that? OK, I will, and you'll see that a strange thing happens, all because the cosine of an angle and the cosine of the negative of that angle are the same. Here's that trig formula:

$$\cos x \ \cos y = \frac{1}{2}\cos(x + y) + \frac{1}{2}\cos(x - y).$$

When we apply it to our sort-of-AM signal, we can get two different results, depending on the order in which we choose x and y:

$$v(t) = 10\cos 2\pi(10^6 + 10^3)t + 10\cos 2\pi(10^6 - 10^3)t \text{ V}$$

or

$$v(t) = 10\cos 2\pi(10^3 + 10^6)t + 10\cos 2\pi(10^3 - 10^6)t \text{ V}.$$

Nyah nyah! you say. See, there are only two frequencies in either of those terms. In the first one the frequencies are 1,001,000 and 999,000 Hz, and in the second one, 1,001,000 and −999,000 Hz. How do you square this with your previous work?

That's easy by looking at the following:

$$10\cos 2\pi(10^3 + 10^6)t = 10\cos 2\pi(-10^3 - 10^6)t.$$

Sure enough, there is −1,001,000 Hz, so all four frequencies are actually present. But this is the hard way—Euler really makes things clearer.

9.3.2 Example of a Spectrum

Let's use as our example of a spectrum the first 14 harmonics of a half-wave-rectified signal. I've chosen to use a 120-V_{rms} 60-Hz source that we get from our standard power receptacles. I put a diode in series with one side of the line and this is the signal that I get:

$$\begin{aligned}
v(t) = \ &54.1 + 85\sin 2\pi 60t - 36.1\cos 2\pi 120t \\
&- 7.2\cos 2\pi 2407 - 3.1\cos 2\pi 360t \\
&- 1.7\cos 2\pi 480t - 1.1\cos 2\pi 600t \\
&- 0.8\cos 2\pi 720t - 0.62\cos 2\pi 840t \text{ V}.
\end{aligned}$$

Fig. 9.10 shows the actual waveform.

In order to get the spectrum of this signal properly plotted, all the terms must be cosine terms, and all amplitudes must be positive. This can be done by simply adjusting phase angles. The sine term becomes a cosine with a phase angle of −90°; the negative amplitudes shift the phase angle by −180°. So the signal, written in a more standard way, becomes

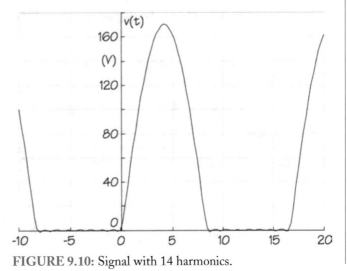

FIGURE 9.10: Signal with 14 harmonics.

$$v(t) = 54.1$$
$$+ 85\cos(2\pi 60t - 90°)$$
$$+ 36.1\cos(2\pi 120t - 180°)$$
$$+ 7.2\cos(2\pi 240t - 180°)$$
$$+ 3.1\cos(2\pi 360t - 180°)$$
$$+ 1.7\cos(2\pi 480t - 180°)$$
$$+ 1.1\cos(2\pi 600t - 180°)$$
$$+ 0.8\cos(2\pi 720t - 180°)$$
$$+ 0.6\cos(2\pi 840t - 180°) \text{ V.}$$

The spectrum for this signal is shown in Fig. 9.11.

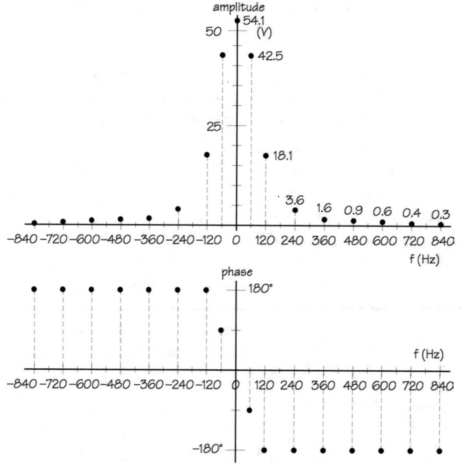

FIGURE 9.11: Discrete spectrum for $v(t)$.

Note the DC term (54.1 V). Note also that there is no phase angle plotted for the DC term (i.e., at zero frequency) because no angle can be defined for DC.

The spectrum shows that the amplitudes decrease as the frequency increases (on the positive side). We can also conclude that the energy available at the various frequencies decreases (because power is proportional to the square of the voltage).

9.3.3 Bandwidth

The *width* of a signal's spectrum can be defined as the signal's *bandwidth*. If we look at the amplitude spectrum in Fig. 9.11, we see that it covers frequencies from zero to 840 Hz. So we would say that this signal has a bandwidth of 840 Hz.

If there were no DC term, we would measure the bandwidth from the lowest frequency to the highest. In our example, if the DC term were absent, the bandwidth would be 840 - 60 = 780 Hz.

Some signals "kinda go on forever" or at least seem to. Then how do we define bandwidth? We can, for example, define the "5% bandwidth" as being the bandwidth figured to where the harmonics are less than 5% of the fundamental. In dB this is $20 \log 0.05 = -26$ dB. We could also define the 2% bandwidth (-34 dB) or the 1% bandwidth (-40 dB) or any other value.

For our signal, the 5% bandwidth is found by looking for terms less than 5% of 85, which is less than 4.25 V. Hence the 5% bandwidth is 240 Hz because the next harmonic has an amplitude of 3.1 V.

What happens if we actually *bandlimit* our signal to, say, the 5% bandwidth? This might happen if we pass the signal through some kind of filter. Fig. 9.12 shows the 5% bandlimited signal plotted with the original signal. Except for some rather small deviations at the bottom and a slight change in the peak, you really can't tell them apart.

9.4   CONTINUOUS SPECTRA

A spectrum doesn't have to be discrete. Discrete spectra apply to signals that are made of individual fixed frequencies. A singer

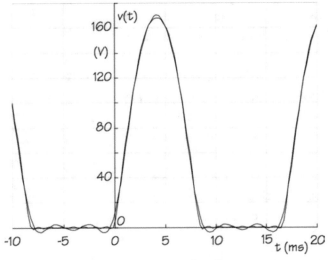

FIGURE 9.12: Bandlimited signal ("5%").

who is singing just one note produces a discrete spectrum (although I doubt that singers think of themselves as spectrum-production apparatus!).

The question of describing, in the frequency domain, what a singer produces when she's singing an aria is an interesting question. Can what she does be described by a discrete spectrum that has dots for all the possible notes? Probably not, especially when the singer glides between notes. So in effect the singer is producing, over time, a continuous spectrum of sounds.

We need continuous spectra when we want to represent, in a general way, the frequencies present in a signal such as singing or speech or FM broadcasting, and so on. What we are saying is that we cannot peg exactly the frequencies present, all the while knowing that we must design amplifiers or filters or transmitters or whatever to process these signals.

Consider a fictitious signal that has a continuous spectrum given by

$$V(f) = \frac{2}{1 + jf} \text{ V/Hz,}$$

which is a *voltage density function*.

The amplitude spectrum for this signal is shown in Fig. 9.13. This is a continuous spectrum that implies that every frequency is present in the signal, including DC (zero frequency). The phase spectrum for the signal is shown in Fig. 9.14.

Another example is the band-limited speech that passes through the telephone system. We might give a simple representation of the spectrum of such a signal as shown in Fig. 9.15.

We are saying that we don't know the exact frequency at any particular time but that we can expect[2] the signal to have any frequency in the range between 300 and 3300 Hz. This gives us essential information if we want to design a system to digitize speech.

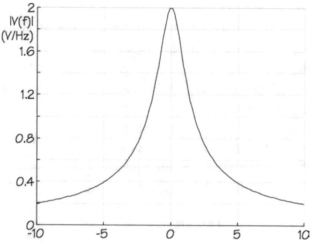

FIGURE 9.13: Continuous amplitude spectrum.

---

[2]This sounds like a probability density function, doesn't it?

9.5   POWER SPECTRA

We are often interested in the power available in a signal, especially the power as a function of frequency. But this is, in a way, a strange question. In Fig. 9.16 for example, how much power is available from that source?

The question is clearly meaningless! I could ask, however, how much power can this source deliver to a 600-Ω resistor as shown in Fig 9.17.

Using v-squared over R, the result is

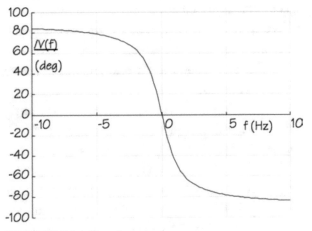

FIGURE 9.14: Continuous phase spectrum.

$$p(t) = \frac{(180\cos 2\pi 60t)^2}{600}$$
$$= 54\cos^2 2\pi 60t \text{ W}.$$

The average power into 600 Ω is

$$P_{AV} = \frac{v_{RMS}^2}{R} = \frac{\left(v_{peak}/\sqrt{2}\right)^2}{R} = 27 \text{ W}.$$

So to be able to answer the question of available power, we need a load. But what load? 600 Ω because it's common in communications? 50 Ω because that's the impedance of a number of different coaxial cables? 3.14159 kΩ because that's 1000 π? Hmmm, none of these seems to make any better sense than another.

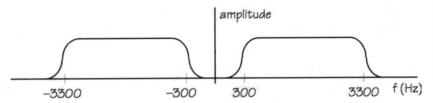

FIGURE 9.15: Uniform spectrum of a phone system.

How about 1 Ω? That makes the division simple! In fact, that's the one we use, and we refer to this as the "1-Ω power" when we want to talk about it. Hence if we have an amplitude spectrum that is calibrated in volts, the 1-Ω power will be one-half the peak amplitude squared. That way, everyone is speaking the same language.

Fig. 9.18 shows a discrete amplitude spectrum, calibrated in volts.

Let's find the 1-Ω power spectrum for this signal. The power values will be the squares of the values. Why squares? Well, power is voltage squared divided by resistance, which is 1 Ω.

But maybe you remember that we must divide by 2 if the voltage is peak rather than rms. True, but here the voltage of a given term is split into two parts, the positive and negative frequency components. So this "2" is already taken care of. The result of this is shown in Fig. 9.19.

How can we get away with adding power? It sure looks like we are applying superposition and we've been told that's a No-No for power. But it works here. Why?

These signal components are all at different frequencies. Products of sinusoids at different frequencies have zero average value. So there are no "cross terms" and we can add the powers directly. (Reread this paragraph so you are sure you don't miss a very important fact!)

Another quantity that we often need is the energy carried by the signal. This makes little sense for discrete spectra, but it is very useful for continuous spectra. With these, we can talk about the amount of energy in some particular band of frequencies.

Go back to the continuous spectrum that I used in the previous section, reproduced here as Fig. 9.20.

We might ask: How much 1-Ω energy is carried by the signal in the band between 5 and

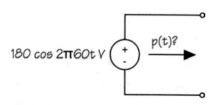

FIGURE 9.16: Source power?

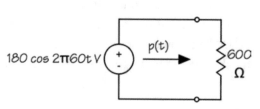

FIGURE 9.17: Source with load.

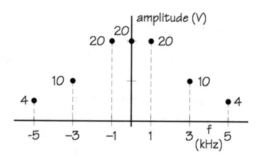

FIGURE 9.18: Amplitude spectrum.

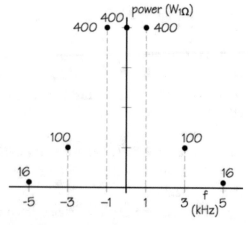

FIGURE 9.19: Power spectrum.

10 Hz? Since the curve is voltage per hertz, we need to square the curve and integrate between 5 and 10 Hz. Don't forget both the $+f$ and $-f$ parts of the spectrum—I doubled the 5-to-10 integral to account for both sides. That integration yields

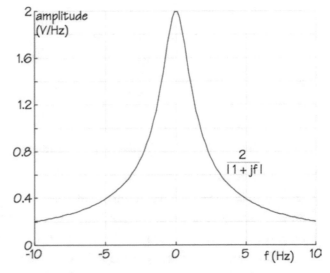

FIGURE 9.20: Amplitude spectrum.

$$W_{1\Omega} = 2\int_5^{10} \left[\frac{2}{|1+jf|} \right]^2 df$$

$$= 2\int_5^{10} \frac{4}{1+f^2} df$$

$$= 0.7818 \text{ W-s.}$$

The result is 782 mW-s (milliwatt-seconds) or 782 mJ (millijoules).

Hmmm. There's a problem that you might have noticed already. When we calculated power for the discrete spectrum of Fig. 9.12, the results were in watts. Here we have integrated a spectrum and gotten watt-seconds. Since integration is just a summation of lots of little parts, what has changed?

The function! This continuous spectrum is a voltage *density* spectrum, or volts per hertz. So we have an extra "seconds" in the result. And that makes it energy, not power.

We can also find the total 1-Ω energy carried by the signal by squaring the curve and integrating over all frequencies (from −∞ to +∞). Since the curve is symmetric, doubling the positive half will do:

$$W_{1\Omega} = 2\int_0^\infty \left[\frac{2}{|1+jf|} \right]^2 df$$

$$= 2\int_0^\infty \frac{4}{1+f^2} df$$

$$= 4\pi = 12.57 \text{ W-s.}$$

Now we can answer questions such as: What bandwidth must we have so that we pass 75% of the total available 1-Ω energy? Let's integrate from 0 to some frequency f_{75} and double the result. Then equate this to 75% of the result above and see what f_{75} turns out to be:

$$W_{1\Omega b} = 2\int_0^{f_{75}} \left[\frac{2}{|1 + jf|} \right]^2 df$$

$$= 8\tan^{-1} f_{75},$$

$$8\tan^{-1} f_{75} = 0.75 \times 4\pi,$$

$$f_{75} = 2.41\,\text{Hz}.$$

Three-quarters of the energy available in this signal is concentrated between 0 and 2.41 Hz.

9.6   DESIGN EXAMPLE

A signal is to be passed through a somewhat noisy channel. Our goal is to get as much of this signal through the channel as possible with the least interference from noise. Let's consider the noise characteristic of the channel and then look at the signal.

The channel is 15 kHz wide. This channel is bothered by white noise, which means that the noise is at the same level for all frequencies (just as white light contains all frequencies). The noise *power* in the channel is shown in the graph in Fig. 9.21. Note that this power is given as a *power density* in μW/Hz.

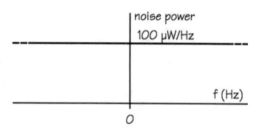

FIGURE 9.21: White noise in channel.

If we want to know how much noise there is in a certain frequency band, we must integrate this noise power over the width of that band. Here, of course, the integration is a simple multiplication. For example, if we want the noise power (really the 1-Ω noise power) in a band from DC to 1 kHz, we calculate

$$noise\ power = 2 \times 100 \times (1000 - 0) = 200\,\text{mW}.$$

Why the factor of 2? The bandwidth must include both the negative and positive halves of the spectrum because we use double-sided spectra.

Now let's look at the signal. Presume this signal is produced by some musical instrument playing middle C (about 512 Hz), but there are some strong overtones. The signal has a fundamental, and then *even* harmonics through the eighth:

$$v_{in} = 20\cos(2\pi 512t + 0°)$$
$$+ 8\cos(2\pi 1024t + 15°)$$
$$+ 6\cos(2\pi 2048t + 40°)$$
$$+ 8\cos(2\pi 3072t + 57°)$$
$$+ 0.5\cos(2\pi 4096t + 69°)\,\text{V}.$$

To get an idea of what this looks like, I've plotted a couple of periods in Fig. 9.22. And after looking at it, I wonder what this instrument sounds like. Maybe we don't want to know! (Actually, I listened to it via Matlab— it's kind of a raucous electric organ.)

The spectrum of the signal and the spectrum of the noise will tell us about what is happening as we work toward our goal. So we need the spectrum of the signal. Since the spectrum is double sided, all the amplitudes are split to appear half on the positive side and half on the negative side. Fig. 9.23 is the amplitude spectrum; the phase spectrum isn't going to be useful in this design problem.

Now comes the hard part. How do we get some measure of how well we are doing? This is often done using the *signal-to-noise ratio*. We write this as *S/N*.

The signal-to-noise ratio is the ratio of the power in the signal to the power in the noise *in the bandwidth of the channel*. It is measured in dB, and since this is a *power* measurement, dB is 10 times the log.

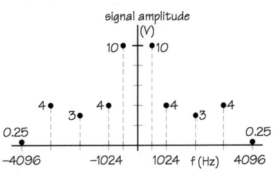

FIGURE 9.22: Two periods of signal.

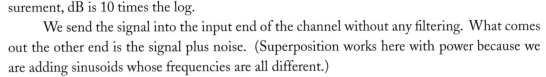

FIGURE 9.23: Signal spectrum.

We send the signal into the input end of the channel without any filtering. What comes out the other end is the signal plus noise. (Superposition works here with power because we are adding sinusoids whose frequencies are all different.)

First let's calculate the amount of power that the signal produces at the output end of the channel. Keep in mind that this is the 1-Ω power. Each component in the amplitude spectrum contributes its amplitude squared in 1-Ω power. So the total signal power coming out of the channel is

$$S_{15} = 0.25^2 + 4^2 + 3^2 + 4^2 + 10^2$$
$$+ 10^2 + 4^2 + 3^2 + 4^2 + 0.25^2$$
$$= 282 \text{ W}_{1\Omega}.$$

Now let's see how much noise power is coming out of the channel (which is 15 kHz wide):

$$N_{15} = 2 \times 15000 \times 100 \times 10^{-6} = 3 \text{ W}.$$

The signal-to-noise ratio for the unfiltered channel is

$$S/N_{15} = 10 \log(S_{15}/N_{15}) = 19.7 \text{ dB}.$$

And just what does this mean? It means that the power of the noise in the channel is only about 20 dB below the power of the signal. That's not too good. You would complain about a telephone line with that S/N ratio; you'd say it was noisy (or lousy!).

What can we do? In this design example I am going to use an ideal low-pass filter, just to keep things fairly simple. The ideal filter has an absolute, 100%, super-perfect cutoff: all frequencies below f_c pass through the filter, and all those above are completely attenuated.

Something that seems to make sense is to filter the output of the channel so that all of the signal passes through the filter but none of the noise at frequencies *beyond* the signal get by. Hence I will set the cutoff frequency f_c of the filter at 4096 Hz. That will just pass the highest frequency in the signal (that eighth harmonic) while chopping off all the noise from 4096 Hz on up to 15 kHz.

What's the S/N ratio now? The signal power remains the same, but the noise power is diminished. The result is

$$N_8 = 2 \times 4096 \times 100 \times 10^{-6} = 819 \text{ mW},$$
$$S/N_8 = 10 \log(S_8/N_8) = 25.4 \text{ dB}.$$

Aha! Things have improved by setting f_c = 4096 Hz. Could we do even better? Yes, you perhaps think, because if we move f_c even further down, we'll cut out more noise. Then you realize that we'll cut out some signal, too.

Cutting out some signal means the signal won't sound the same. Perhaps we can tolerate that. But we must take into account that the signal *power* is reduced when we chop off a harmonic.

OK, let's do this—let's set f_c at 3072 Hz, keeping harmonics through the sixth and chopping off the eighth. Here are the calculations:

$$S_6 = 4^2 + 3^2 + 4^2 + 10^2$$
$$+ 10^2 + 4^2 + 3^2 + 4^2$$
$$= 282 \text{ W}_{1\Omega}.$$
$$N_6 = 2 \times 3072 \times 100 \times 10^{-6} = 614 \text{ mW}.$$
$$S/N_6 = 10 \log(S_6/N_6) = 26.6 \text{ dB}.$$

Gosh, a modest improvement! If we keep the eighth harmonic, $S/N = 25.4$ dB; if we chop it off, $S/N = 26.6$ dB. But is 1.2 dB worth it?

But let's continue and chop off the sixth harmonic also. The signal power is reduced further, and so is the noise. If we set f_c at 2048 Hz and calculate S/N as before, we get

$$S/N_4 = 10\log(S_4/N_4) = 27.9 \text{ dB}.$$

Hmmm, still better, so let's keep going. If I make the same calculations for $f_c = 1024$ Hz (keeping the second harmonic) and $f_c = 512$ Hz (keeping only the fundamental), I get

$$S/N_2 = 10\log(S_2/N_2) = 30.5 \text{ dB},$$
$$S/N_1 = 10\log(S_1/N_1) = 32.9 \text{ dB}.$$

Each of these is a little better. If we keep only the fundamental, $S/N = 32.9$ dB, and that's 7.5 dB better than keeping everything through the eighth harmonic. Is this enough of a difference to be meaningful, at least in the realm of signal-to-noise ratios? Hard to say. In fact, what we have here is a good engineering dilemma: Do we literally destroy the sound of the instrument to achieve this improvement?

I'd choose to set $f_c = 4096$ Hz and keep the entire signal while chopping off all the rest of the noise above that. After all, the true sound of an instrument requires us to hear the overtones. Fig. 9.24 shows the final design.

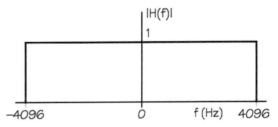

FIGURE 9.24: Filter for signal.

Realize, of course, that no ideal filter can exist. But the things we did here apply equally to problems using realizable filters, although in a more complicated way.

9.7   SUMMARY

In this chapter we've seen one way to represent a signal in the frequency domain, namely, by showing the signal's spectrum. This spectrum provides the magnitude and the phase of each component of a signal in graphical form.

Our spectra are double sided, meaning that we draw both the positive and the negative halves of the frequency axis. In reality, we know that only positive frequencies exist, but the double-sided artifice keeps the mathematics correct for complicated signals such as those involving modulation. The numerical value of the amplitude of a certain sinusoidal term in a signal is split 50-50 between the negative and the positive frequency components.

The "numbers" that inhabit a drawing of a spectrum also lead directly to phasors that describe the signal. Except for a factor of 2, the amplitude value and the phase angle in a spectrum give the phasor description of the signal at that frequency.

We will discover another connection in the next chapter when we take up the Fourier series in exponential form. The numbers that come out of this transformation are also phasors, the same ones, in fact, that we have been dealing with in this chapter.

CHAPTER 10

Fourier Series: Period

How can we describe signals that are periodic but are more complicated than simple sinusoids? Waveshapes like triangular waves and sawteeth and things with funny bumps in them? You certainly weren't expecting that the answer would be *Fourier series*, were you?

Fourier series are used to represent *periodic* waveforms. In themselves, they aren't very complicated. The surprise comes when you realize that they connect very nicely with two things we have already dealt with: phasors and superposition.

10.1 FOURIER SERIES MATH

A Fourier series can be written for just about any *periodic* function. There are mathematical restrictions on what functions can be represented this way. We'll leave those for mathematicians to worry about because all signals generated by real physical systems meet the restrictions.

There are two common forms of Fourier series. One is based on sines and cosines and is usually called the *trigonometric series*. The other is based on exponentials and is called the *exponential series*. By now, we've used Euler's formula enough, so we do know a way to convert from one to the other.

But we won't. We are going to deal exclusively with the exponential form of the series. Why? Well first, the coefficients of the exponential form are also the coefficients that we use for frequency spectra. No manipulation of any kind is needed. The magnitudes of the complex coefficients produce the amplitude spectrum, while their phase angles produce the phase spectrum.

Second, the coefficients of the exponential form of the series are phasors, pure and simple. So once we have the coefficients, we have the phasors for each of the frequencies present in the signal. Neatly set up for the kind of analysis we've already learned! (Wellllll, I lie a little—there's that factor of 2 that we must remember to include.)

Off we go into the math of the exponential form.

10.1.1 Exponential Fourier Series

The exponential form of the Fourier series represents a *periodic* time function in the following way:

$$g(t) = \sum_{i=-\infty}^{\infty} c_i e^{j 2\pi i f_o t},$$

where c_i is the *complex* amplitude of the ith term, f_o is the fundamental frequency of the periodic waveform (and hence 1/period), and $e^{j\cdots}$ is the exponential representation of a sinusoid (with some help from Euler).

Now let's play around with this a bit. First, we can always split a summation into a number of individual pieces. I'll split this one into negative and positive values of the index i:

$$g(t) = \sum_{i=-\infty}^{-1} c_i e^{j 2\pi i f_o t} + \sum_{i=0}^{\infty} c_i e^{j 2\pi i f_o t}.$$

Let's improve the symmetry by pulling out the $i = 0$ term:

$$g(t) = \sum_{i=-\infty}^{-1} c_i e^{j 2\pi i f_o t} + c_o + \sum_{i=0}^{\infty} c_i e^{j 2\pi i f_o t}.$$

Change the index i in the first term to $-i$, which changes the signs of the limits and of the exponent:

$$g(t) = \sum_{i=\infty}^{1} c_{-i} e^{-j 2\pi i f_o t} + c_o + \sum_{i=1}^{\infty} c_i e^{j 2\pi i f_o t}.$$

But summations can be done in any order, so this can be rearranged still further:

$$g(t) = \sum_{i=1}^{\infty} c_{-i} e^{-j 2\pi i f_o t} + c_o + \sum_{i=1}^{\infty} c_i e^{j 2\pi i f_o t}.$$

Now where are we? Well, we shouldn't be too surprised to find that the coefficients c_i are complex. If we don't believe that right now, we'll see that it is true when we do some examples.

Since the coefficients are complex, they *must* appear in complex pairs, one the complex conjugate of the other. Otherwise, $g(t)$ won't be real. So it should be no surprise that the coefficients c_{-i} are the complex conjugates of the coefficients c_i. These coefficients have magnitude and phase.

The coefficients in the first summation (the c_{-i} terms) are the spectrum (magnitude and phase) to the left of zero frequency. The c_o term is the DC term. The coefficients of the second summation (c_i) are the spectrum to the right of zero frequency.

OK, that's all very fine. But how do we find the coefficients themselves? As you might expect, there is a way that isn't hard to do, especially if you have a "math doer" like Mathematica or Maple.

Recall that the Fourier series works only for periodic functions. Hence any function we consider must have a period—call this T_o. This period is related to the fundamental frequency f_o by $f_o = 1/T_o$.

The basic formula for finding the coefficients is

$$c_i = \frac{1}{period} \int_{period} f(t)e^{-j2\pi i f_o t}\, dt.$$

We can pick the alignment of the "period" in that integral just about anywhere. For example, we can start at $t = 0$, giving

$$c_i = \frac{1}{T_o} \int_0^{T_o} f(t)e^{-j2\pi i f_o t}\, dt.$$

Let's choose a period that is symmetric about the origin:

$$c_i = \frac{1}{T_o} \int_{-T_o/2}^{T_o/2} f(t)e^{-j2\pi i f_o t}\, dt.$$

Or we can start somewhere else, as long as we include exactly one period.

And that's all there is to it! Granted that this integral may require some effort. The result may be an interesting mess, too, especially from the "math doers." We'll see some of this in the examples that follow.

10.2   EXAMPLES
Let's start with a very simple case and then deal with several more interesting signals.

10.2.1 Single Sinewave
The simplest case I can think of (except for DC, which is trivial) is a single sinusoid:

$$v(t) = 10\cos 2\pi 1000t \text{ V}.$$

We'll convert this to exponential form so that we can clearly see the numbers that we must use to draw the spectrum:

$$v(t) = 10\frac{e^{j2\pi 1000t} + e^{-j2\pi 1000t}}{2}$$

$$= 5e^{j2\pi 1000t} + 5e^{-j2\pi 1000t} \text{ V}.$$

This last expression *is* the Fourier series for $v(t)$. I'll say it again, "This last expression *is* the Fourier series for $v(t)$." That's important! The first "5" is the value of c_1; the second is the value of c_{-1}.

So the amplitude spectrum has a dot at "5" at frequencies of $+1000$ Hz and -1000 Hz. The phase spectrum has dots for $0°$ at those frequencies. Fig. 10.1 shows the results graphically.

Now I'll try to get this same result using the formula for the Fourier series. First, we need the period. The fundamental (and only) frequency is 1000 Hz, so the period is going to be $T_o = 1/1000 = 10^{-3}$ s.

The integration is

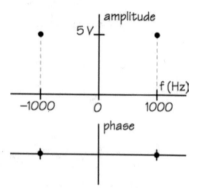

FIGURE 10.1: Simple spectrum.

$$c_i = \frac{1}{10^{-3}} \int_0^{10^{-3}} 10\cos 2\pi 1000t \; e^{-j2\pi i 1000t} \, dt.$$

The result, at least as I got it from Maple, is a bit of a mess, even after using the simplify operation:

$$c_i = 5\frac{j\left(ie^{-2j\pi i} - i\right)}{\pi i^2 - \pi}.$$

By hand I casn simplify this a little more:

$$c_i = \frac{j5i}{\pi\left(i^2 - 1\right)}\left(e^{-j2\pi i} - 1\right).$$

If you substitute various values of i into this, you quickly discover that $c_i = 0$ for all i except for $i = 1$. There it blows up because of the denominator. (Actually, when you try values like $i = 2$ in

Maple, you get numerical values, but they are very small, caused most likely by round-off errors in evaluating the exponential.)

So what to do about $i = 1$? Redo the integration with i specifically set to 1 and you'll get

$$c_1 = 5.$$

Doing the same thing for $i = -1$ also yields 5.

That gives us everything we need to know about the spectrum. First, only c_1 and c_{-1} have values. Second, those values are real, or to say it another way, they are complex with phase angles of zero. The Fourier series is therefore

$$v(t) = 5e^{j2\pi1000t} + 5e^{-j2\pi1000t} \text{ V}.$$

which matches what we got before.

Wow! All that math to find out that our result matches the spectrum we already knew we had! But now we can tackle more complicated signals.

10.2.2 Half-wave Rectified Sine Wave

Consider the signal shown in Fig. 10.2, a half-wave rectified sine wave (well, really a cosine). Find the Fourier coefficients for this.

Its period is 40 ms, so the fundamental frequency is $f_o = 1/40$ ms = 25 Hz.

The function is

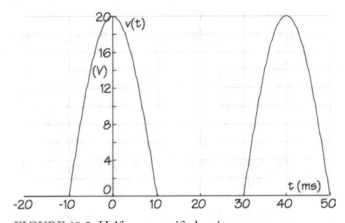

FIGURE 10.2: Half-wave rectified cosine wave.

$$v(t) = \begin{cases} 20\cos 2\pi25t \text{ V} & \text{for } -10 \text{ ms} < t < 10 \text{ ms} \\ 0 & \text{for } 10 \text{ ms} < t < 30 \text{ ms.} \end{cases}$$

The integration for the coefficients is

$$c_i = \frac{1}{40\times10^{-3}} \int_{-10\times10^{-3}}^{10\times10^{-3}} 20\cos 2\pi25t\, e^{-j2\pi i 25t}\, dt.$$

Maple simplified the result to

$$c_i = -20\frac{\cos(\pi i/2)}{\pi(i^2-1)}.$$

Here again, the $i = 1$ case requires a separate integration.

The result from Maple, carried out to $i = 14$, looks like this:

$c_0 = 6.366$ $c_5 = -0.43 \times 10^{-6}$ $c_{10} = 0.064$

$c_1 = 5.0$ $c_6 = 0.182$ $c_{11} = 0.32 \times 10^{-5}$

$c_2 = 2.122$ $c_7 = -0.34 \times 10^{-5}$ $c_{12} = -0.045$

$c_3 = -0.81 \times 10^{-6}$ $c_8 = -0.101$ $c_{13} = -0.20 \times 10^{-5}$

$c_4 = -0.424$ $c_9 = 0.26 \times 10^{-5}$ $c_{14} = 0.033$.

Coefficients like c_3 are artifacts, caused by round-off in Maple. They should be zero. You have to watch for such things when you trust a "math doer."

Note that some of the coefficients are negative. As phasors, these are positive magnitudes with phase angles of $-180°$. This means that the similar term for negative i should have a phase angle of $+180°$. (I chose $-180°$ for the positive side because real systems always delay a signal. In other words, they can't predict.)

So the meaningful coefficients for this half-wave rectified sine wave are

$$c_0 = 6.366 \underline{/0°} \text{ V}$$

$c_{-1} = 5.0 \underline{/0°}$ V $c_1 = 5.0 \underline{/0°}$ V

$c_{-2} = 2.122 \underline{/0°}$ V $c_2 = 2.122 \underline{/0°}$ V

$c_{-4} = 0.424 \underline{/180°}$ V $c_4 = 0.424 \underline{/-180°}$ V

$c_{-6} = 0.182 \underline{/0°}$ V $c_6 = 0.182 \underline{/0°}$ V

$c_{-8} = 0.101 \underline{/180°}$ V $c_8 = 0.101 \underline{/-180°}$ V

$c_{-10} = 0.064 \underline{/0°}$ V $c_{10} = 0.064 \underline{/0°}$ V.

I cut this off when the magnitudes of the coefficients were smaller than 1% of the fundamental c_1.

10.2.3 An Important Pulse Train

The Fourier series for a train of pulses is important for two reasons. First, we often work with such pulse trains. Second, this leads to the Fourier transform of Chapter 12. Right now, though, it's just an exercise in finding coefficients.

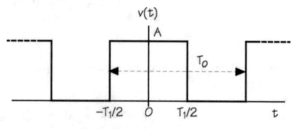

FIGURE 10.3: General pulse train.

Fig. 10.3 shows the general pulse train. The amplitude is A, the period is T_o, and the width of the pulse (its "on" time) is T_1.

The integration is not difficult but the result is messy and, in Maple at least, takes some manipulation to achieve a decent result. I've used $1/T_o$ instead of f_o in the exponent.

$$c_i = \frac{1}{T_o} \int_{-T_1/2}^{T_1/2} A e^{-j2\pi it/T_o}\, dt$$

$$= \frac{A}{\pi i} \sin\left(\pi i \frac{T_1}{T_o}\right)$$

$$= \frac{AT_1}{T_o} \frac{\sin\left(\pi i {T_1}/{T_o}\right)}{\pi i {T_1}/{T_o}}.$$

The last equation is in the form of $(\sin x)\,/\,x$, something that we will see again. All of the coefficients are real. Some have a minus sign, so the phase angles are either $0°$ or $\pm 180°$.

It is interesting to plot the size of the terms c_i as a function of i. To do this, I need numbers, so I have chosen to make $A = 1$ and $T_1 = T_o/5$. Fig 10.4 shows a continuous plot of c_i versus i. We must realize, of course, that the c_i exist only for integer values of i. We also realize that parts of the curve that are negative really represent positive coefficients whose phase angles are $\pm 180°$.

This curve is going to show up again in Chapter 12.

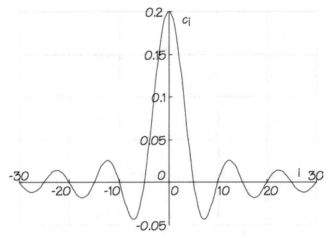

FIGURE 10.4: "$\sin x/x$" values of c_i for $A=1$ and $T_1=T_o/5$.

10.2.4 Filtered Triangular Wave

Now we can use our newly acquired knowledge to find out what happens to a triangular wave as it passes through a particular filter.

For the filter I'll use the second-order active circuit from Section 7.4, reproduced here as Fig. 10.5. The transfer function $H(s)$ of this filter is

$$H(s) = \frac{2,800,000}{s^2 + 200s + 1,000,000}.$$

I want to find out what happens when the triangular wave of Fig. 10.6 passes through this filter. While the job is not difficult, it is a bit tedious. We will find the coefficients c_i of the Fourier series for the waveform. These we know are phasors. We'll apply superposition and pass each one individually through the filter at the appropriate value of frequency.

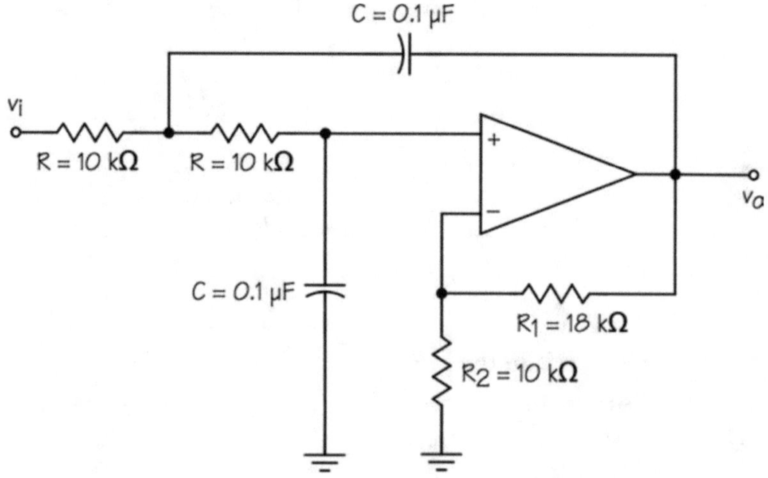

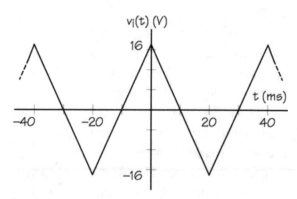

FIGURE 10.5: Second-order filter.

First, we'll find the coefficients of the Fourier series of the input. The period is 40 ms. I split the function into two pieces, the upslope to the left of $t = 0$ and the downslope to the right of $t = 0$. (Because the integral for c_i has $e^{-j\cdots}$ in every term, the integration is not symmetric about the origin, so I have to do the whole thing.)

The input is

FIGURE 10.6: Triangular input signal.

$$v(t) = \begin{cases} 1600(t + 0.01) & \text{for } -10 \text{ ms} < t < 0 \\ -1600(t - 0.01) & \text{for } 0 < t < 10 \text{ ms}. \end{cases}$$

The integration is as follows, where c_o has to be done separately:

$$c_i = \frac{1}{0.04} \int_{-0.02}^{0} 1600(t+0.01)\, e^{-j2\pi it/0.04}\, dt$$

$$+ \frac{1}{0.04} \int_{0}^{0.02} -1600(t-0.01)\, e^{-j2\pi it/0.04}\, dt,$$

$$c_o = \frac{1}{0.04} \int_{-0.02}^{0} 1600(t+0.01)\, dt$$

$$+ \frac{1}{0.04} \int_{0}^{0.02} -1600(t-0.01)\, dt.$$

The result of all this, after some head scratching and simplifying (and remembering that $e^{j\pi i}$ is either +1 for even values of i or −1 for odd values of i), is

$$c_i = \begin{cases} 0 & \text{for i even} \\ \dfrac{64}{\pi^2 i^2} & \text{for i odd.} \end{cases}$$

Curiously, all terms are positive and have 0° phase angles.

How many of these terms do I take? We have to cut off the series somewhere. Considering that most of our circuit components are precise to only about 5%, we can stop fairly early. I have chosen to cut off the series when the coefficients get smaller than 1% of the fundamental c_1. That means I keep terms through c_9. The result is

$c_1 = 6.485$ at $f_1 = f_o = 1/T_o = 25$ Hz,
$c_3 = 0.720$ at $f_3 = 3f_o = 75$ Hz,
$c_5 = 0.259$ at $f_5 = 5f_o = 125$ Hz,
$c_7 = 0.132$ at $f_7 = 7f_o = 175$ Hz,
$c_9 = 0.080$ at $f_9 = 9f_o = 225$ Hz.

Fig. 10.7 is a plot of the input using just these terms to the 1% cutoff. It looks pretty good!

Now I use these coefficients as phasors. But there's something that is easy to forget: a *factor of* 2. The coefficient c_1 is 6.485, so the phasor this represents is 12.970 $\underline{/0°}$ V. (Recall that you can justify, at least in your mind, the need for 2 by saying you are including both the positive and the negative frequencies of the spectrum.)

The calculation to get the filter's output V_{o1} for the phasor given to us by the coefficient c_1 is

$$V_{o1} = 2c_1 H(s)\big|_{s=j2\pi f_1}$$

$$= 12.970\angle 0° \frac{2800000}{(j2\pi 25)^2 + 200(j2\pi 25) + 1000000}$$

$$= 37.212\angle -2° \text{ V}.$$

Similar calculations at the remaining frequencies give us the result of passing this signal through the filter:

$$\begin{aligned}
V_{o1} &= 37.212 \angle\underline{-2}° \text{ V} &&\text{at } f_1 = f_o = 25 \text{ Hz}, \\
V_{o3} &= 5.149 \angle\underline{-7}° \text{ V} &&\text{at } f_3 = 3f_o = 75 \text{ Hz}, \\
V_{o5} &= 3.508 \angle\underline{-22}° \text{ V} &&\text{at } f_5 = 5f_o = 125 \text{ Hz}, \\
V_{o7} &= 2.442 \angle\underline{-134}° \text{ V} &&\text{at } f_7 = 7f_o = 175 \text{ Hz}, \\
V_{o9} &= 0.432 \angle\underline{-164}° \text{ V} &&\text{at } f_9 = 9f_o = 225 \text{ Hz}.
\end{aligned}$$

These phasors can be converted to the time domain in the usual way. The resultant $v_o(t)$ is shown below, with a plot in Fig. 10.8:

$$\begin{aligned}
v_o(t) = \ &37.212\cos(2\pi 25t - 2°) \\
&+5.149\cos(2\pi 75t - 7°) \\
&+3.508\cos(2\pi 125t - 22°) \\
&+2.442\cos(2\pi 175t - 134°) \\
&+0.432\cos(2\pi 225t - 164°) \text{ V}.
\end{aligned}$$

Uhuh, I know what you are thinking—so what! I do admit that the filter didn't really do anything much of interest, so this must be written off as an academic exercise in using the Fourier series and working with phasors.

But just in case the spectra might be interesting, they are shown in Fig. 10.9, using dots for the input and plusses for the output. It does show some considerable changes wrought by the filter. You might want to look back to Fig. 7.19 to see the filter's characteristics. (Keep in mind that the axis in that figure in Chapter 7 is in *rad/s*, not hertz.)

Other than an overall gain, the biggest changes are in the phase

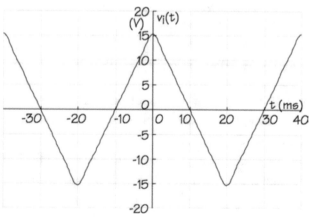

FIGURE 10.7: Fourier-series input to 1%.

spectrum, where the phase delay grows quickly with increasing frequency.

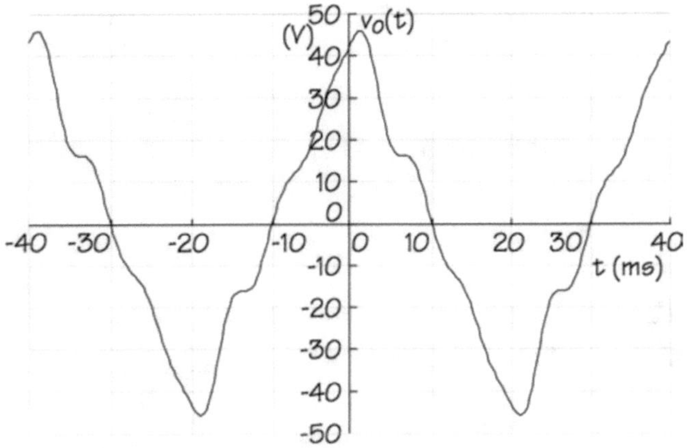

10.3   DESIGN EXAMPLE

Create the waveform shown in Fig. 10.10, including as many terms as are needed to get the result to within 2% of the design goal.

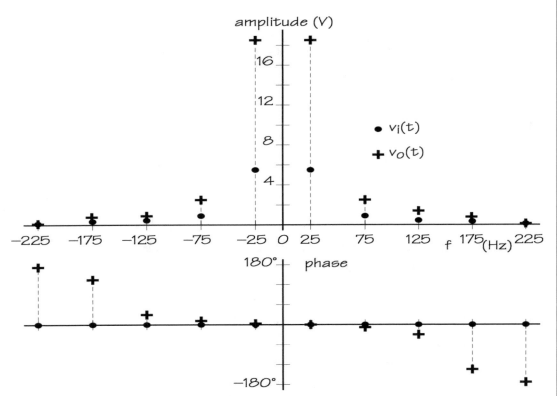

FIGURE 10.8: Output due to Fourier input.

Let's start by finding the Fourier exponential series coefficients for the given waveform. The period is 20 µs and the waveform has four sections:

FIGURE 10.9: Input and output spectra for filter.

$$v_a = \frac{2,500,000}{3} t \text{ V},$$

$$v_b = 5 \text{ V},$$

$$v_c = -\frac{2,500,000}{3} t + \frac{40}{3} \text{ V},$$

$$v_d = 0 \text{ V}.$$

The three nonzero functions are applied to the c_i integrals to get

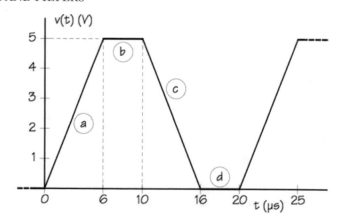

FIGURE 10.10: Signal to synthesize.

$$c_i = \frac{1}{20 \times 10^{-6}} \left[\int_0^{6 \times 10^{-6}} \frac{2500000}{3} t\, e^{-j2\pi it/20 \times 10^{-6}}\, dt \right.$$

$$+ \int_{6 \times 10^{-6}}^{10 \times 10^{-6}} 5 e^{-j2\pi it/20 \times 10^{-6}}\, dt$$

$$\left. + \int_{10 \times 10^{-6}}^{16 \times 10^{-6}} \left(-\frac{2500000}{3} t + \frac{40}{3} \right) e^{-j2\pi it/20 \times 10^{-6}}\, dt \right].$$

The result is a bit of a mess, as we often expect. Moreover, the $i = 0$ term must be found via a separate integration:

$$c_o = 2.5,$$

$$c_i = \frac{25}{6\pi^2 i^2} \left[e^{-\frac{3}{5} j\pi i} + e^{-j\pi i} - e^{-\frac{8}{5} j\pi i} - 1 \right].$$

A check of the e-to-the stuff shows that terms with an even index i are zero. The results through $i = 9$, using $f_o = 1/20$ μs $= 50$ kHz, are

$c_o = 2.50$ at DC,
$c_1 = -1.105 - j0.803 = 1.366 \underline{/-144°}$ at 50 kHz,
$c_3 = -0.018 + j0.055 = 0.058 \underline{/-252°}$ at 150 kHz,
$c_5 = -0.068 - j0 = 0.068 \underline{/-180°}$ at 250 kHz,
$c_7 = -0.003 - j0.010 = 0.011 \underline{/-108°}$ at 350 kHz,
$c_9 = -0.014 + j0.010 = 0.017 \underline{/-216°}$ at 450 kHz.

In these results, I have chosen all the angles to be negative, simply because all circuits delay signals. The values for the c_i with negative i are the same except that the phase angles will be positive.

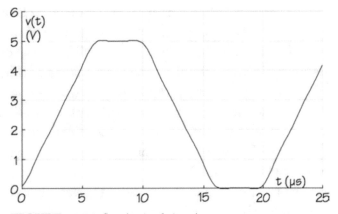

FIGURE 10.11: Synthesized signal.

Why did I go to the $i = 9$ term? After all, 2% of 2.5 is 0.025 and both c_7 and c_9 are beneath that cutoff. Yes, but the specifications said that we were to create a result that was within 2% of the given waveform.

That leaves a question of how much does each term contribute to improving the result. We should suspect that just saying that a coefficient is less than 2% of the largest coefficient probably does not guarantee what we are really looking for. So a little engineering judgment says to use an extra term or so. (One way to check is to plot the difference between the design waveform and the Fourier result.)

The proof of whether this works is to design and construct a circuit to do the job. I'm not going to do that here, but Fig. 10.11 shows that the waveform generated by these terms at least looks like it does the job.

Remember that, to create the time-domain result from the c_i, I must include a factor of 2 in the amplitudes represented by the phasors. The resultant waveform is

$$
\begin{aligned}
v(t) = {} &2.5 \\
&+2.732\cos(2\pi 50000t - 144°) \\
&+0.116\cos(2\pi 150000t - 252°) \\
&+0.135\cos(2\pi 250000t - 180°) \\
&+0.021\cos(2\pi 350000t - 108°) \\
&+0.034\cos(2\pi 450000t - 216°) \text{ V.}
\end{aligned}
$$

10.4 SUMMARY

The Fourier series is a summation of terms that represent any periodic waveform. The series is a "perfect" representation if we include all the terms, even if this means going to infinity. But for most of our common signals, the magnitudes of the terms fall off fairly rapidly as the frequency increases. Discarding higher frequency terms often does little to distort the representation of the signal.

The coefficients of the exponential form of the Fourier series are *phasors*, including magnitude and phase. We must double the magnitude of any amplitude coefficient to get the traditional phasor. The magnitudes of the coefficients lead directly to the double-sided amplitude spectrum of a signal; the phase angles provide the phase spectrum.

Since the Fourier representation of a signal is a sum of terms of different frequencies, we can apply superposition when we want to pass such a signal through some circuit. This works even if we want to combine *power* terms. (Recall that power calculations involve products of sine terms.) Products are not linear operation, so in general we can't apply superposition to power. But the average value of the product of two sine terms that have different frequencies is zero, so average power applies to each frequency individually. Each term's effect, either voltage or power, can be analyzed separately and the results added together at the end.

We started out in this chapter by pointing out several times that the Fourier series is for *periodic* functions. What happens if they are not? We'll take that up in Chapter 12. In the meantime, Chapter 11 will deal with the design of some very important circuits called *filters* that alter, hopefully in a useful way, the amplitude and phase spectra of signals.

CHAPTER 11

Filter Design: By the Book

What's a filter? The round can in your car that the oil goes through to…. Actually, the analogy isn't too bad. The oil filter alters the fluid that is flowing through it, "improving" it in some way. The electrical filter does the same thing, altering the signal that flows through it, "improving" it in some way.

We usually describe filters in the phasor or frequency domain. We'll say something like, "Design a filter that will pass only those frequencies between 20 and 800 Hz." Or, "Design a filter to cut out frequencies near 40 kHz." These filters change the character of the signal to accomplish some goal.

In this chapter we are going to take a look at some general types of filters and a few of their uses. We'll study the characteristics of a simple "unit" filter. We are going to deal here only with *active filters*, which are those that include amplifiers, generally op-amps.

Filter design can get complicated and can quickly go beyond the level where we want to focus in this course. So while we'll design filters such as Butterworth, we are going to do most of this "by the book." We'll stick to standard forms for which the analysis has already been done. And pretty much stay with low-pass filters, too.

11.1 FILTER BASICS

Consider a "perfect" filter, one that we know we can't possibly build. The "perfect" filter is absolutely unattainable, but we can come as close to it as we have money to spend on the effort.

There are four fundamental types of filters, described in general terms by what they do to the signal:

> *Low-pass* filters pass only those frequencies below a certain cutoff frequency. The pass band for an ideal low-pass filter is shown in Fig. 11.1. Note that I have drawn both posi-

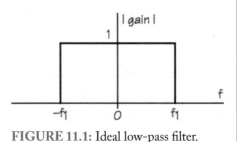

FIGURE 11.1: Ideal low-pass filter.

tive and negative frequencies. Usually, however, we look at just the positive side. When we talk about a low-pass filter, we say that it passes frequencies below the frequency f_1.

A low-pass filter is used in CD players to filter the audio signal after it is converted from digital form. The cutoff frequency is about 20 kHz and prevents artifact, caused by the digital conversion, that would sound horrible.

High-pass filters pass frequencies above a certain cutoff frequency. Fig. 11.2 shows this. Here again, we say that it passes frequencies above the frequency f_2. A high-pass filter might be installed on the antenna input to a TV set to keep out the interference from a local CBer. In this case the cutoff would be set to about 50 MHz.

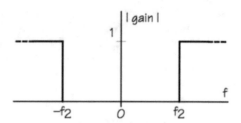

FIGURE 11.2: Ideal high-pass filter.

Band-pass filters allow a range of frequencies to pass through as shown in Fig. 11.3. It cuts off those frequencies outside the passband. A band-pass filter might be used in an interface with the telephone system to restrict the incoming signal to the voice range of 300 to 3300 Hz.

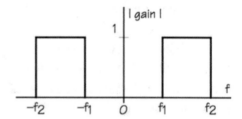

FIGURE 11.3: Ideal band-pass filter.

Band-stop or *notch* filters cut out a particular range of frequencies. Fig. 11.4 shows this. The filter passes all frequencies except those between f_1 and f_2. A common use is in AM receivers to cut out the 10-kHz whistle that can develop.

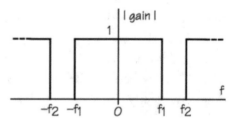

FIGURE 11.4: Ideal band-stop filter.

These filters are all ideal. In other words, we can build them only if we have an infinite amount of money.

11.1.1 Simple Low-Pass Active Filter

Let's start by looking at the circuit for a simple filter. The filter shown in Fig. 11.5 is an active filter. There are several things that we can analyze:

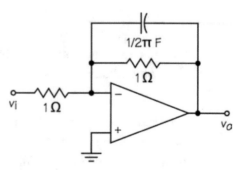

FIGURE 11.5: Unit low-pass active filter.

- gain versus frequency, remembering that "gain" generally implies the magnitude of the output divided by the input, often in dB;
- phase angle versus frequency, usually in degrees; and
- step response, which is a time-domain analysis that we'll leave to homework or laboratory.

If we remember that the transfer function for an op-amp circuit of this type (inverter) is the ngative of the feedback impedance divided by the input impedance, we can arrive at $H(s)$ easily:

$$Z_f(s) = \frac{1 \times \dfrac{1}{(1/2\pi)s}}{1 + \dfrac{1}{(1/2\pi)s}},$$

$$Z_{in}(s) = 1,$$

$$H(s) = -\frac{Z_f(s)}{Z_{in}(s)} = \frac{-1}{s/2\pi + 1}.$$

Most of the time, though, we are interested in what filters do as a function of f in hertz rather than s:

$$H(f) = \frac{-1}{(j2\pi f)/2\pi + 1} = \frac{-1}{1 + jf}.$$

(Now you can see why I chose the strange value for the capacitor—it makes $H(f)$ come out in very neat form.)

What should we do with this result? One thing to do is to look at the gain of the filter as a function of frequency. "Gain" means the "magnitude of $H(f)$." Fig. 11.6 shows the gain over two decades of frequency. This is, of course, the Bode diagram that you already know about.

We call 1 Hz the *cutoff frequency* because this is the frequency at which the curve makes a clear turn. Do you recall how many dB below the flat part of the curve we

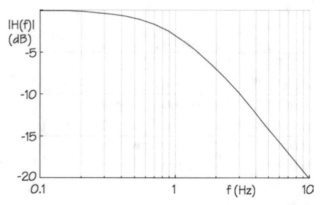

FIGURE 11.6: $|H(f)|$ of unit low-pass filter.

should be at the cutoff frequency? Yes, the graph shows 3 dB, a number that we almost always use for this purpose.

The frequency $f = 1$ Hz is sometimes called the *corner frequency* because it is where the "corner" of the curve lies. It is also the *half-power point*, because −3 dB in voltage is $1/\sqrt{2}$ and power is voltage squared and therefore 1/2.

We can talk about the number of poles that the function $H(s)$ has. A pole is located where the denominator goes to zero. Here it would be at $s/2\pi + 1 = 0$, or $s = -2\pi$ s^{-1}. This filter has a single pole.

The curve has two rather clear *asymptotes*. The one on the left is the *passband* of the filter and is at 0 dB. The second is the downward slope on the right. Convince yourself with a straight edge that this slope is −20 dB per decade of frequency. This slope is called the *rolloff* of the filter.

We saw in Chapter 7 that a circuit's response in the frequency domain often showed this −20 dB/decade slope or an integer multiple of it. This is simply a characteristic of *all* circuits using the components we have, and in fact of all linear systems. This means that if someone wants us to design a filter with a slope of, say, −30 dB/decade, we simply cannot do it.

Recall that 20 dB/decade is the same as 6 dB/octave, where an octave is a factor of 2 in frequency.

The phase response of our filter is shown in Fig. 11.7. Note that it passes through 135° at the cutoff frequency. The asymptotes of this curve are 180° on the left and 90° on the right, although we have to go a decade left or right to be able to see these asymptotes.

The phase response shown includes the negative sign of the gain. Without that, the plot would go from 0° to −90°. This represents a *delay* of the signal. For example, at the cutoff frequency of 1 Hz, one full cycle of 360° takes 1 s. Hence a phase angle of −45° means that the signal is delayed by 45/360 = 0.125 s.

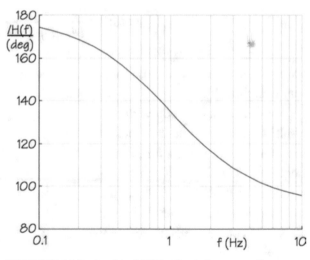

FIGURE 11.7: Angle of $H(f)$ of unit low-pass filter.

This delay is not uniform with frequency. If $f = 0.1$ Hz, the phase angle is −5.7°. One cycle takes 10 s, so the delay is −5.7/360 x 10 = 0.159 s. At $f = 10$ Hz, a similar calculation yields a delay of 0.023 s. A signal passing through this filter is delayed more at lower frequencies than at higher ones.

So what do we have in the circuit of Fig. 11.5? A low-pass filter with a cutoff frequency of 1 Hz, a passband of 0 dB, and a rolloff of 20 dB/decade. Moreover, it is a "unit filter" in the sense that the elements are either "1" or adjusted to make the cutoff frequency equal to 1.

11.1.2 Building Blocks

The low-pass filter that we just worked with (Fig. 11.5) is a "standard" low-pass filter. It has a *unit* frequency response because the cutoff frequency is 1 Hz. We often use such unit circuits as building blocks. We start with such a unit filter and then make three adjustments:

Cutoff frequency scaling. We can scale the cutoff frequency by adjusting the circuit component values to get a different value. We do this by noting what happens to impedance as a function of frequency: We want the impedances to remain the same as we alter the frequency to a new value.

- The impedance of a resistor is R, which is not a function of frequency. Hence changing the cutoff frequency to a new value, say, double the original unit frequency, will have no effect on the values of any resistors in the circuit.
- The impedance of an inductor is $j2\pi fL$. If we, say, double the frequency f, we must halve the inductance L to keep the same impedance.
- The impedance of a capacitor is $1/j2\pi fC$. If we, say, double the frequency f, we must halve the capacitance C to keep the same impedance.

What all of this says is that if we wish to move the cutoff frequency from f_1 to kf_1, we must change any inductor value to L/k and any capacitor value to C/k.

Input impedance. Our standard unit filter has an input impedance of 1 Ω, rather low! We often have restrictions on this because of the loading effect of our filter on the circuit that is driving it. But we must make this change carefully or we will alter the other characteristics of the filter.

In Fig. 11.5, suppose we want an input impedance of 1 kΩ. That says we should change the 1-Ω input resistor to 1 kΩ. That's fine, except that it changes the gain of the filter in the passband. It's easy to see how much by noting that, in the passband, the frequency is well below $f = 1$ Hz. In this region, the capacitor looks like an open circuit. So the gain of the circuit before our change is $-1/1 = -1$. When we change the input impedance to 1 kΩ, the gain goes down to $-1/1000 = -0.001$.

We must keep the altered gain in mind when we make the last adjustment, that of the passband gain.

Passband adjustment. If we want a passband gain of something other than 0 dB, we must add either amplification or attenuation. Often we can use the op-amp already in the circuit to do this. This can be a confusing step, but if you think carefully about what you are doing, you'll get it right.

In the low-pass filter's passband, capacitors generally are not a factor. For frequencies well below the cutoff frequency, capacitors look like open circuits. For frequencies well above, they look like short circuits. Hence the passband gain is often controlled by the ratio of the feedback and the input resistors.

But there's a catch! If we simply alter resistors without thinking about the capacitors, we'll move the cutoff frequency. For either the series or the parallel combination of an R and a C, the $1/RC$ product determines the position of the pole and hence the cutoff frequency. So if we, say, increase the value of R to adjust the gain, we must decrease the value of C by the same ratio.

The best way to see how this is done is through an example. (Well, the very best way is to do it yourself several times!)

11.1.3 Adjusting Our Filter

Take the filter of Fig. 11.5 and design a low-pass filter circuit that will provide a 6-dB gain in the passband, have a cutoff frequency of 2 kHz, and have an input imped-ance of 600 Ω. Use commercially available components. The Bode diagram of Fig. 11.8 shows our design goal asymptotically.

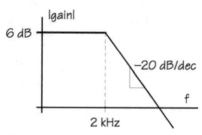

FIGURE 11.8: Design goal.

1. First, we move the cutoff frequency from 1 Hz to 2 kHz. We have increased the frequency by a factor of 2000, so the value of the capacitor must be reduced by the same factor. The new $C = 1/(2\pi 2000)$ F.

2. Next, we adjust the input impedance to 600 Ω without changing the passband gain or the cutoff frequency. I must increase the 1-Ω input resistor by a factor of 600. There-fore I must also increase the value of the feedback resistor by a factor of 600 if I want the passband gain to stay the same.

3. At this point, both resistors are 600 Ω. But I have altered the RC product of the feedback impedance, so I must also adjust C to account for this. Since R went up by a factor of 600, C must go down by the same factor. So now $C = 1 / [(2\pi 2000)(600)] = 0.1326$ μF.

4. Finally, we increase the passband gain from 0 dB to 6 dB. We calculate "dB in reverse" to get the actual change:

$$gain = 10^{dB/20} = 10^{6/20} = 2.$$

(We should have known that 6 dB is a factor of 2!)

To increase the gain by a factor of 2 requires me to change the feedback resistor by that factor. I can't change the input resistor without altering the input impedance. Instead, I change the feedback resistor from 600 Ω to 1200 Ω.

5. But this change also changes the RC product. R went up by a factor of 2, so C must come down by the same factor. Finally, $C = 0.1326 / 2 = 0.0663$ μF.

6. Reviewing the results:

$$R_{in} = 600\,\Omega,$$
$$R_f = 1200\,\Omega,$$
$$C = 0.0663\,\mu F.$$

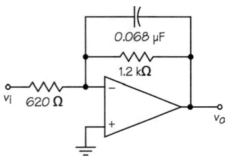

FIGURE 11.9: Final design.

7. Now we choose commercial values. R_f already is, because a 1.2-kΩ resistor is available. The nearest capacitor is 0.068 μF, a slightly larger value by 2.5%. This will lower the cutoff frequency by 2.5% to about 1950 Hz. That's probably OK. Finally, for R_{in} we'll use the closest resistor value of 620 Ω. That will make our passband gain 1200/620 = 5.7 dB. That's probably OK, too.

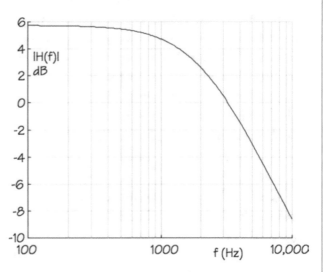

FIGURE 11.10: Final result.

The final circuit design is shown in Fig. 11.9 and the frequency response is shown in Fig. 11.10.

You might check the Bode diagram carefully to see if the cutoff frequency is really at 2 kHz, or at least close to that. And does the passband gain look right?

11.1.4 High-Pass and Bandpass Filters

While we aren't going to spend a lot of time on filters other than low-pass ones, we'd better at least mention the others. Our unit low-pass filter can be converted to a high-pass filter by just moving the capacitor to a position in series with the input resistor.

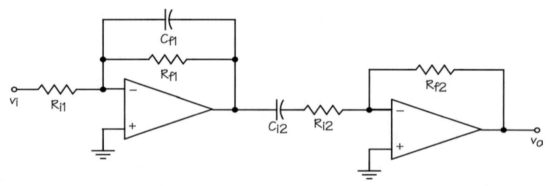

FIGURE 11.11: Bandpass filter using low-pass and high-pass filters.

Bandpass filters are a little trickier, mainly because the obvious design is not the best one. It would seem that we could make a bandpass filter by combining low-pass and high-pass filters. This will work, but it takes two op-amps.

Fig. 11.11 shows this. In this circuit I have placed the low-pass filter first, for no particular reason. The high-pass filter is on the right; notice the capacitor in series with the input resistor.

But a better bandpass filter can be built with just one op-amp as shown in Fig. 11.12. This circuit combines low-pass and high-pass elements into a circuit built around one op-amp. The feedback impedance provides the upper cutoff frequency f_2 (i.e., *low*-pass) at $1/(2\pi R_f C_f)$ Hz, while the input impedance provides the lower cutoff frequency f_1 at $1/(2\pi R_i C_i)$ Hz.

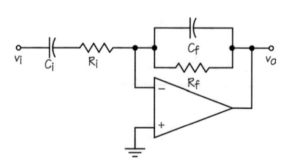

FIGURE 11.12: Better bandpass filter.

11.2   GENERAL DESIGNS

When we are designing circuits involving op-amps, there just aren't too many places to put impedances around these op-amps. We commonly add the condition that we want the op-amp to be the output of the circuit to provide stage-to-stage isolation of loads.

There are three simple circuits that we need to consider. The first, the inverter that we have already used, is shown in Fig. 11.13. Here the gain of the circuit is simply related to the ratio of the impedances. Its input impedance is Z_{in}.

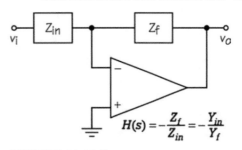

$$H(s) = -\frac{Z_f}{Z_{in}} = -\frac{Y_{in}}{Y_f}$$

FIGURE 11.13: Inverter.

The noninverter is also very useful (Fig. 11.14), mainly because its input impedance is extremely high.

The divider with gain, shown in Fig. 11.15, is easy to use because the "gain" part of the circuit (the noninverter) and the "frequency" part of the circuit (the voltage divider) are separate. The gain is controlled by the resistors, while the "filter" characteristic is controlled by the divider impedances.

Let's go to a design example to see how these are used.

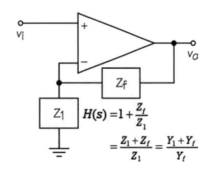

$$H(s) = 1 + \frac{Z_f}{Z_1}$$
$$= \frac{Z_1 + Z_f}{Z_1} = \frac{Y_1 + Y_f}{Y_f}$$

FIGURE 11.14: Noninverter.

11.3 FILTER EXAMPLE

Perhaps the best way to illustrate the use of the model filters and Bode diagrams is to design a filter. So our problem is to design a filter whose frequency-domain characteristic is given by the asymptotes in the Bode diagram of Fig. 11.16 and whose input impedance is at least 1 kΩ at DC.

Before we start, we need to make reasonably sure that the filter can actually be built. One simple test is to see whether the slopes are integer multiples of 20 dB per decade (or 6 dB per octave).

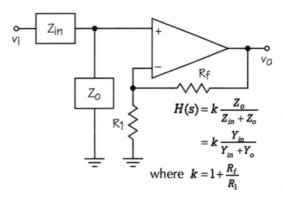

$$H(s) = k \frac{Z_o}{Z_{in} + Z_o}$$
$$= k \frac{Y_{in}}{Y_{in} + Y_o}$$
where $k = 1 + \frac{R_f}{R_1}$

FIGURE 11.15: Divider with gain.

In the Bode diagram, the two asymptotes look fine, with a low-frequency gain of −6 dB and a high-frequency gain of 20 dB.

How about the slope between those two? The frequency from one end of the slope to the other changes by a factor of 20. That means the gain should change by a factor of…hmmmm, how much?

Well, if the frequency changed by a factor of 10 (that's a decade), the gain should change by 20 dB. And if the frequency changed by a factor of 2 (that's an octave), the gain should change by 6 dB. Since gain in dB is logarithmic, I can add the two. Hence a frequency change of 20 leads to a gain change of 26 dB. Yup, that's what's happening, so we should be able to realize this slope.

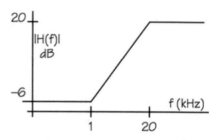

FIGURE 11.16: Design example.

The next step is to write the transfer function from inspection of the Bode diagram. We can collect three important values from the diagram:

- There must be a zero at $f = 1$ kHz because the curve turns upward at that corner frequency. (We know this can't be a double zero because the slope changes from 0 dB/decade to 20 dB/decade, not some multiple of 20.) Hence the transfer function must have in its numerator the factor $s + 2\pi10^3$.
- There must be a pole at $f = 20$ kHz because the curve turns downward by 20 dB/decade (from 20 dB/decade upward to zero slope) at that corner frequency. Hence the transfer function must have in its denominator the factor $s + 2\pi20 \times 10^3$.
- For frequencies much less than 1 kHz (i.e., down to DC), the gain must be -6 dB $= 0.5$.

Now I can write $H(s)$, but with an arbitrary gain that I'll get right in the next step:

$$H(s) = A\frac{s + 2\pi10^3}{s + 2\pi20 \times 10^3}.$$

I get A here by setting s to 0 and the magnitude of $H(0)$ to 0.5:

$$|H(0)| = 0.5 = A\frac{2\pi10^3}{2\pi20 \times 10^3} = \frac{A}{20},$$

$$A = 10, \text{ so}$$

$$H(s) = 10\frac{s + 2\pi10^3}{s + 2\pi20 \times 10^3}.$$

It is useful to have the numerator and denominator factors in the form $s/a + 1$. So I'll rewrite $H(s)$ by dividing through, both in the numerator and the denominator, by the coefficient of s^0:

$$H(s) = 0.5\frac{\dfrac{s}{2\pi10^3} + 1}{\dfrac{s}{40\pi10^3} + 1}.$$

Does this look like any of the "standard" circuits shown in Section 11.2? Well, $H(s)$ is a simple ratio of two polynomials, and, except for a minus sign, the inverter could be a circuit for this. The inverter's transfer function is

$$H(s) = -\frac{Z_f}{Z_{in}}.$$

To match either the numerator or the denominator of the desired $H(s)$ we need to find a circuit that has an impedance function $s/a + 1$. If we can find one, then the numerator polynomial will yield Z_f and the denominator will yield Z_{in}.

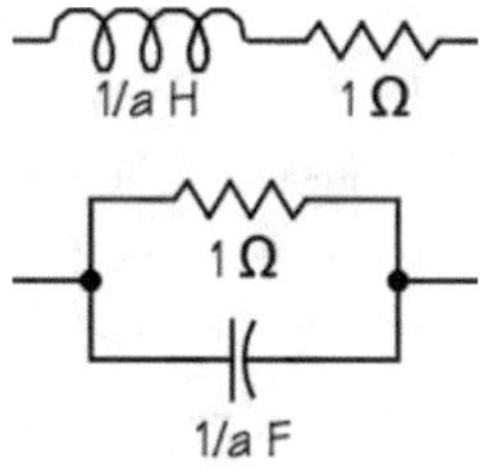

Fig. 11.17 shows a possible circuit for an impedance $s/a + 1$. An inductor whose value is $1/a$ henries has an impedance of s/a. Put this in series with a 1-Ω resistor and we have $s/a + 1$. So this says we could build the numerator with a $1/2\pi10^3$-henry inductor in series with a 1-Ω resistor. The denominator would be similar.

FIGURE 11.17: $(1+s/a)$.

But there's a catch! We prefer to avoid inductors where we can. They are expensive elements and generally have significant losses that alter our designs. Capacitors are usually better choices. But a capacitor in series with a resistor doesn't provide the right form of impedance.

Let's try parallel as shown in Fig. 11.17. The impedance is

$$Z = \frac{1\dfrac{1}{s/a}}{1+\dfrac{1}{s/a}} = \frac{1}{\dfrac{s}{a}+1}, \text{ so}$$

$$Y = \frac{s}{a} + 1.$$

Aha! The admittance has the right polynomial form and uses a capacitor. If we look back at the standard form of the inverter in Fig. 11.13, we find it also has an admittance form:

$$H(s) = -\frac{Y_{in}}{Y_f}.$$

So I will use two parallel RC circuits to get two admittances:

$$H(s) = -\frac{Y_{in} = \dfrac{s}{2\pi10^3}+1}{Y_f = \dfrac{s}{40\pi10^3}+1}.$$

Fig. 11.18 shows the circuit to implement this design. But the design has three problems. First, the element values aren't usually values we can use with standard op-amps, since resistors should generally be "a few kΩ to a few tens of kΩ." Second, the input impedance does not meet our specification of a DC input impedance of at least 1 kΩ (at DC it's now only 1 Ω). Third, the DC gain is not the correct value (it's 1 now).

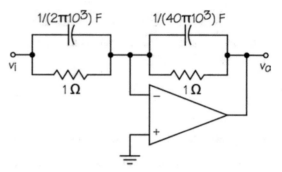

FIGURE 11.18: Initial "unit" design.

Let's fix the DC input impedance first. At DC, the capacitors are open circuits, so the input impedance at DC is 1 Ω. We'll change this to 1 kΩ by changing the input resistor to 1 kΩ. That will require our changing the feedback resistor to 1 kΩ to keep the overall gain the same.

But changing a resistor in an RC combination requires us to adjust the capacitor to keep the RC product constant. Otherwise we move the pole or the zero improperly. So since I multiplied the resistors by 1000, I

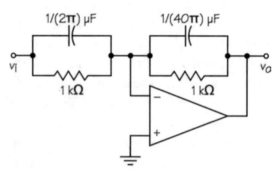

FIGURE 11.19: Design after Z scaling.

must divide the capacitors by 1000. Fig. 11.19 shows the design after this step.

Note that I have also solved the problem of the wrong range of values for op-amp circuits (resistors are now "a few k").

Now we need to deal with the DC gain. Since I need a gain of 0.5, I could change the input resistor to 2 kΩ. Or I could change the feedback resistor to 500 Ω. But what I will really do is keep in mind that their ratio must be 0.5 and work on the problem of "real" components.

The input capacitor at the moment is 0.159 μF, not a standard value. I could change this downward to 0.1 μF or upward to 0.22 μF. I chose (after a couple of tries) to go downward, so the input capacitor is 0.1 μF.

That means I must adjust the input resistor upward to keep the RC product constant. Since C went down by a factor of 1.59, R must go up by that factor. Hence it becomes 1.59 kΩ, which I'll convert to the commercial value of 1.6 kΩ. Note that the DC input impedance is now 1.6 kΩ, which meets specifications.

Now adjust the feedback resistor to get the DC gain correct. This resistor must therefore be 1.6 kΩ/2, which is 800 Ω. I'll use the commercial value of 820 Ω, an error of about 2.5%.

But that change means I must adjust the feedback capacitor to keep that *RC* product constant. The original resistor was 1000 Ω; the new one is 820 Ω. The result is

$$\frac{1000}{820} \times \frac{1}{40\pi} \mu F = 0.0097 \to 0.01 \, \mu F.$$

Whew! The final design is shown in Fig. 11.20 and a plot of the actual gain versus frequency is in Fig. 11.21. The design looks acceptable. Check it against specs!

11.4 SECOND-ORDER DESIGN

Second-order circuits are particularly neat because they can provide two poles with only one op-amp. Their poles can be complex, too, whereas the models we have seen so far can't do this. So second-order circuits give us a wider range of models for filter design.

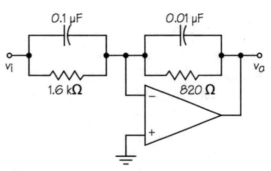

FIGURE 11.20: Final design.

The second-order low-pass filter is shown in Fig. 11.22. Note that both resistors have the same value, as do both capacitors. The gain of the noninverting op-amp is an important part of the second-order character of the circuit.

The second-order high-pass filter is very similar and is shown in Fig. 11.23. The only change is the exchange of the *R*s and the *C*s. Here again, the gain of the op-amp is important in our design.

We'll use the second-order low-pass model as a building block in the next section. And we'll leave some interesting questions about these circuits to homework problems.

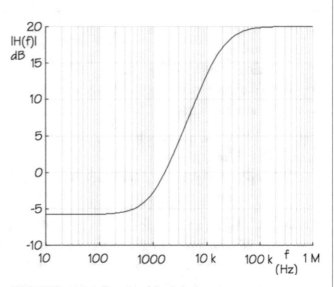

FIGURE 11.21: Result of final design.

11.5   GENERAL FILTER DESIGN

What we are going to do in this section is the handbook design of low-pass filters. Handbook design means that we look at model circuits to find one that has the right shapes of its characteristics, then adjust the circuit elements to get the filter we want.

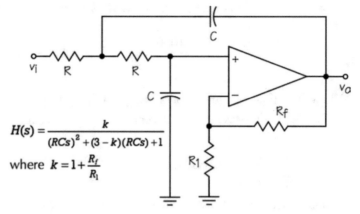

$$H(s) = \frac{k}{(RCs)^2 + (3-k)(RCs) + 1}$$

where $k = 1 + \dfrac{R_f}{R_1}$

FIGURE 11.22: Second-orders low-pass filter.

We will consider only low-pass filters because you will learn the basics of handbook filter design with just this one case. High-pass filter design is done the same way except that, wherever s appears, you write $1/s$ instead. Bandpass design, which can be done by combining low- and high-pass filters, can be done in more advanced ways that yield less expensive results.

General specifications for a filter usually tell us a number of specific things:

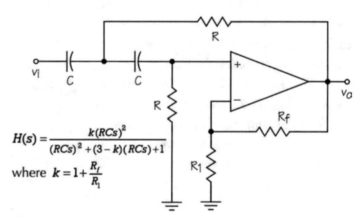

$$H(s) = \frac{k(RCs)^2}{(RCs)^2 + (3-k)(RCs) + 1}$$

where $k = 1 + \dfrac{R_f}{R_1}$

FIGURE 11.23: Second-order high-pass filter.

- Passband gain, usually specified to be within a range of 3 dB, but sometimes less.
- Cutoff frequency.
- Stopband response or rolloff, usually stated either as "x dB per decade" or "at least y dB down at the frequency f."

There are numerous different forms that our design can take. We are going to restrict our discussion (and the examples) to just three: cascaded first order, Butterworth, and Chebychev.

In each of the examples the specifications will be the same. We are to design a low-pass filter that has a passband gain of 12 dB, a cutoff frequency of 20 kHz, and a rolloff so that the output is 45 dB below the passband at 100 kHz. We are also to use standard commercial components.

11.5.1 Cascaded First Order

Our building block for cascading first-order filters will be an *RC* network followed by a noninverter to provide the needed gain and, most important, to isolate one circuit from the next. Fig. 11.24 shows the building block.

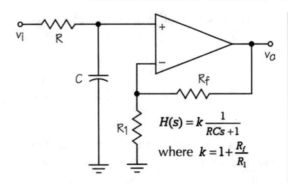

$$H(s) = k\frac{1}{RCs+1}$$

$$\text{where } k = 1 + \frac{R_f}{R_1}$$

FIGURE 11.24: First-order low-pass filter.

Note that the output of the *RC* "voltage divider" is connected to the high-impedance input of the op-amp. Hence the op-amp has little effect on the characteristic of the *RC* network.

We will design our filter by cascading these for as many stages as are required to get the right rolloff. Fig. 11.25 shows this. All stages are identical. We can cascade these building blocks because the op-amp makes each stage independent.

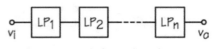

FIGURE 11.25: Cascading filters.

Cascading first-order circuits gives a greater and greater slope in the rolloff part of the specification. One stage gives us 20 dB per decade. Two stages give us 40. Three stages give 60, and so on.

Our building block has the following characteristic:

$$H(s) = k\frac{1}{RCs+1},$$

$$\text{where } k = 1 + \frac{R_f}{R_1}.$$

The general form for a low-pass filter is

$$H(s) = k\frac{1}{\dfrac{s}{2\pi f_c} + 1},$$

so this tells us how to choose *R* and *C*:

$$RC = \frac{1}{2\pi f_c},$$

where f_c is the cutoff frequency *of this stage*.

When we cascade n such identical blocks, we get

$$H_{total}(s) = \left[\frac{k}{s/2\pi f_c + 1} \right]^n.$$

But there's a catch that is easy to miss. If one stage has a gain of −3 dB at the cutoff frequency, two cascaded stages will have a gain of −6 dB. So n stages cascaded will have a gain of −3n dB at the cutoff frequency. That won't do, because we generally expect only −3 dB overall.

To correct this problem, we must move the cutoff frequency of each stage to a *higher* value. The adjustment is

$$f_c = \frac{f_{cutoff}}{\sqrt{2^{1/n} - 1}},$$

where f_{cutoff} is the specified cutoff frequency of the filter we are designing and f_c is the cutoff frequency that we design into each stage. f_c is larger than f_{cutoff}.

There are several ways to figure out how many stages are required to accomplish the specified rolloff, but the easiest to use is graphical. (Yes, really, even in this age of computers.) The graph of Fig. 11.26 is a plot of the gain of cascaded first-order filters for a *normalized* cutoff frequency.

Now let's see how to do all this by designing the filter that we have already described.

The graph will give us the order of the filter n. In our case, we require the cutoff frequency to be 20 kHz and the output to be 45 dB below the passband at 100 kHz. This means that the output is to be 45 dB below the passband at five times the cutoff frequency, namely, $5f_c$.

Find on the graph the point where f/f_c = 5 and where the output equals −45 dB. Select the curve that will give at least −45 dB. It appears that n = 10 will do this. The n = 9 curve looks pretty close, but we generally need to be conservative here.

So now we know that we need to cascade at least nine first-order stages to design a filter that will meet our specifications. I'll add one stage to give me a slight margin. Just for the record, that's ten op-amps!

We select the actual cutoff frequency f_c for each stage by using

$$f_c = \frac{20 \times 10^3}{\sqrt{2^{1/10} - 1}} = 74.655 \text{ kHz}.$$

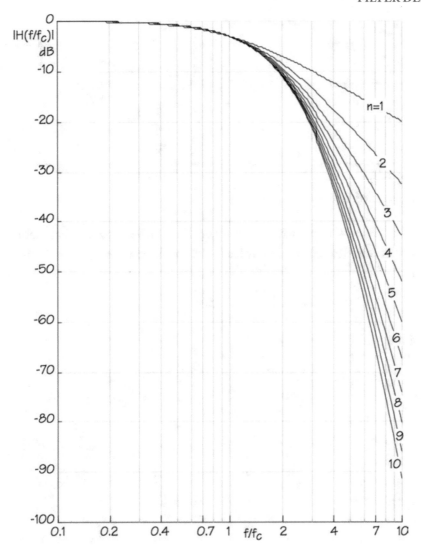

FIGURE 11.26: Normalized curves for a cascaded first-order filter.

Now use the relationship between the RC product and the cutoff frequency f_c to get values for R and C:

$$RC = \frac{1}{2\pi 74.655 \times 10^3} = 2.132 \times 10^{-6},$$

$$R = \frac{2.132 \times 10^{-6}}{C} = \frac{2.132}{C\,\mu\text{F}}.$$

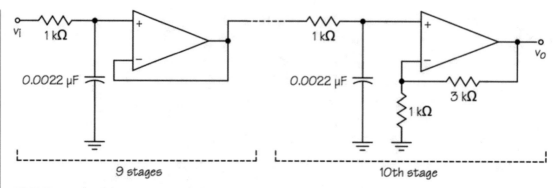

FIGURE 11.27: Ten-stage cascaded first-order design.

I tried several commercial values of C to see what R would be. I not only looked for reasonably close commercial values of R but also for an R of at least 1 kΩ to keep proper values around the op-amp. My result is

$$C = 0.0022\,\mu\text{F},\ R = 1\,\text{k}\Omega.$$

I need an overall gain of only 12 dB, which is a factor of 4 (well, really 3.98 but...), so I choose to use a gain of 1 for each stage except the last. There I will use a 3-kΩ resistor and a 1-kΩ resistor to provide the gain of 4.

This circuit has nine identical stages, all with a gain of 1, and a tenth stage with a gain of 4. All use the same values of R and C. Fig. 11.27 shows the circuit, and Fig. 11.28 shows the frequency-domain plot of the filter's response. The dotted lines mark the 3-dB-down value of the corner frequency and the −45-dB-down point at five times the corner frequency. Both look acceptable.

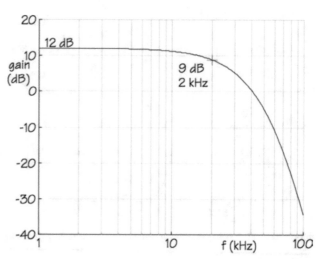

FIGURE 11.28: Ten-stage cascaded first-order design.

11.5.2 Cascaded Second-Order Design

The second-order filter models that we saw in Section 11.4 can be cascaded to achieve somewhat better results. Here, "better" means a circuit with fewer op-amps. We'll leave that to the problems and go on with an even better design.

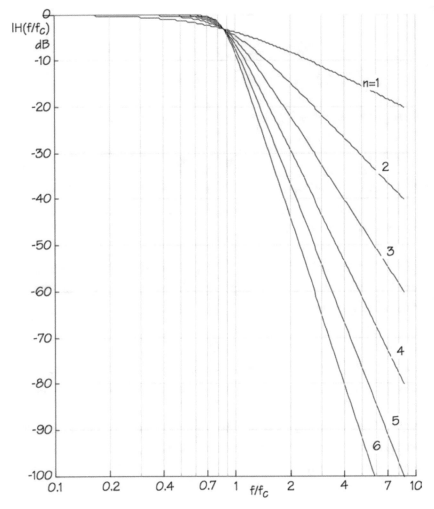

FIGURE 11.29: Normalized curves for Butterworth filter.

11.5.3 Butterworth Design

Mr. Butterworth[1] proposed a design many years ago that will allow us to build filters that generally use fewer op-amps than the cascaded first-order design requires. In fact, he did a lot of the work for us so that we really will do a handbook design this time.

The Butterworth filter places the poles of the transfer function on a circle around the origin. This provides a filter that is called *maximally flat*, meaning that the passband has no wiggles in it. It requires fewer stages for any particular rolloff requirement than the cascaded first-order design requires.

[1]Probably not related to Mrs. Butterworth of squeeze-bottle-spread fame.

n=1	$(s/2\pi f_c)+1$
2	$(s/2\pi f_c)^2 +1.414(s/2\pi f_c)+1$
3	$\left[(s/2\pi f_c)+1\right]\left[(s/2\pi f_c)^2 +(s/2\pi f_c)+1\right]$
4	$\left[(s/2\pi f_c)^2 +0.7654(s/2\pi f_c)+1\right]\left[(s/2\pi f_c)^2 +1.848(s/2\pi f_c)+1\right]$
5	$\left[(s/2\pi f_c)+1\right]\left[(s/2\pi f_c)^2 +0.6180(s/2\pi f_c)+1\right]\left[(s/2\pi f_c)^2 +1.618(s/2\pi f_c)+1\right]$
6	$\left[(s/2\pi f_c)^2 +0.5176(s/2\pi f_c)+1\right]\left[(s/2\pi f_c)^2 +1.414(s/2\pi f_c)+1\right]\left[(s/2\pi f_c)^2 +1.932(s/2\pi f_c)+1\right]$

FIGURE 11.30: Butterworth denominators.

How is this possible? Butterworth filters start their downward plunge at a frequency below the cutoff frequency, yet still have a gain of −3 dB at the cutoff frequency. This "steeper plunge" means the curve is already down more than we can achieve using cascaded first-order design.

To say this another way, Mr. Butterworth was clever!

As before, we find the needed number of stages from a graph of normalized Butterworth gains. That graph is shown in Fig. 11.29. Note that the slopes are all still integer multiples of −20 dB/decade.

Let's see where we fall with the design example we started in the previous section. We want to be 45 dB below the passband at five times the cutoff frequency. Looking at the graph in Fig. 11.29, we see that we can do this with $n = 4$. ($n = 3$ falls a bit short.) Gosh, that's a great improvement over $n = 10$, isn't it?

The polynomials for the Butterworth design are already tabulated for us. Fig. 11.30 gives the first six of these. One of them will be used as the denominator of the $H(s)$ for our filter design.

We choose the polynomial for $n = 4$. But to do what? Well, look at the polynomial and note that it is the product

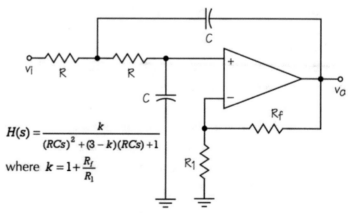

$$H(s) = \frac{k}{(RCs)^2 +(3-k)(RCs)+1}$$

where $k = 1 + \dfrac{R_f}{R_1}$

FIGURE 11.31: Second-order low-pass filter.

of two second-order polynomials. We already know about second-order low-pass filters. Fig. 11.31 reproduces the building block that we saw in Section 11.4.

I will use two of these cascaded. One will implement the first half of the fourth-order Butterworth denominator; the other will do the second half.

First, write the left half of the Butterworth denominator with f_c replaced by 20 kHz. Then compare it with the denominator for $H(s)$ for the second-order low-pass building block. This will give us values for RC and and hence for R_f/R_1. Finally, select workable commercial values for these components.

Here's the work for the first half of our filter:

$$\left[\left(\frac{s}{2\pi 20\times 10^3}\right)^2 + 0.7654\left(\frac{s}{2\pi 20\times 10^3}\right)+1\right]$$

$$\downarrow$$

$$(RCs)^2 + (3-k)(RCs)+1,$$

$$\text{so } RC = \frac{1}{2\pi 20\times 10^3} = 7.9577\times 10^{-6}$$

$$\text{and } 3-k=0.7654, k=2.2346$$

$$\text{so } \frac{R_f}{R_1} = k-1 = 1.2346.$$

I tried several different combinations of values before arriving at

$$R = 3.6\,\text{k}\Omega, C = 0.0022\,\mu\text{F},$$
$$R_f = 1.6\,\text{k}\Omega, R_1 = 1.3\,\text{k}\Omega.$$

Now do the same thing for the right half of the fourth-order Butterworth polynomial. But this time we don't have to deal with RC because that value stays the same. In fact, because each stage of a Butterworth filter has the same RC value, this type of filter is said to be *synchronously tuned.*

$$\left[\left(\frac{s}{2\pi 20\times 10^3}\right)^2 + 1.848\left(\frac{s}{2\pi 20\times 10^3}\right)+1\right]$$

$$\downarrow$$

$$(RCs)^2 + (3-k)(RCs)+1,$$

$$\text{so } 3-k=1.848, k=1.152$$

$$\text{so } \frac{R_f}{R_1} = k-1 = 0.152,$$

$$R_f = 3\,\text{k}\Omega, R_1 = 20\,\text{k}\Omega.$$

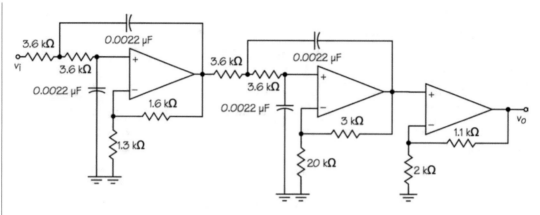

FIGURE 11.32: Butterworth design.

Whew, you say, that does it. Nope! What about the overall gain of the two stages together? We are supposed to have 12 dB in the passband. But each stage provides a gain of only k (see the $H(s)$ in Fig. 11.31 if you don't remember this). Hence we must add a stage of gain.

The gains of the stages are

$$k_1 = 1 + \frac{1.6}{1.3}, k_2 = 1 + \frac{3}{20},$$
so $k_{total} = 2.565 = 8.18$ dB.

We need 12 dB, which is 3.98, so we need to add gain:

$$k_{added} = \frac{3.98}{2.565} = 1.552.$$

I'll use another noninverting stage for this, so $R_f = 0.552R_1$. Values that work are

$$R_f = 1.1\,\text{k}\Omega, R_1 = 2\,\text{k}\Omega.$$

Now put all this together, namely, two second-order stages and one stage to adjust the overall gain. Fig. 11.32 shows the final circuit, and Fig. 11.33 is the plot to verify that this works.

11.5.4 Chebychev Design

Mr. Chebychev was also a clever person. His design scheme is a little different from Mr. Butterworth's. Chebychev's poles are on an ellipse. His filter responses are by no means flat. Instead, they are *equal ripple* in the passband, but the ripples are limited to some design amount. (We will use a 3-dB limit here.)

Chebychev filters also do not have the same RC values for each stage, so they are called *stagger tuned*, for whatever that is worth.

The result of all this is that Chebychev filters start their plunge even more violently. That means that they can have a somewhat sharper rolloff than is possible with Butterworth filters. Hence we sometimes will need fewer stages to do the same job.

Don't forget, though, that we still can have slopes of only $20n$ dB/ decade. No other values are possible.

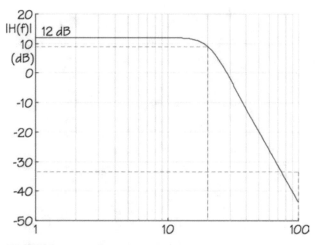

FIGURE 11.33: Butterworth design.

Chebychev design is done the same way that Butterworth design is done. We need the graph of Fig. 11.34 to figure out the number of stages. We need the Chebychev denominator polynomials in Fig. 11.35 to select the proper component values.

Now consider our design example again. We are to be 45 dB below the passband at five times the cutoff frequency. A look at Fig. 11.34 shows that this can be done with $n = 3$. So I'll select the $n = 3$ Chebychev denominator from Fig. 11.35.

That denominator has one first-order term and one second-order term. My filter is going to have a first-order stage just like that of Fig. 11.24 (the one we used for cascaded first-order design). It's also going to have a second-order stage just as we did for the Butterworth filter (Fig. 11.31).

Let's do the second-order match-up to Fig. 11.31 first, selecting the second term of the Chebychev polynomial, replacing f_c with 20 kHz, and matching this to the building block:

$$\left[\left(\frac{s}{2\pi 20 \times 10^3 \times 0.9159}\right)^2 + 0.3245\left(\frac{s}{2\pi 20 \times 10^3 \times 0.9159}\right) + 1\right]$$

$$\downarrow$$

$$(RCs)^2 + (3 - k)(RCs) + 1,$$

$$\text{so } RC = \frac{1}{2\pi 20 \times 10^3 \times 0.9159} = 8.6884 \times 10^{-6},$$

$$\text{and } 3 - k = 0.3245, \ k = 2.6755$$

$$\text{so } \frac{R_f}{R_1} = k - 1 = 1.6755.$$

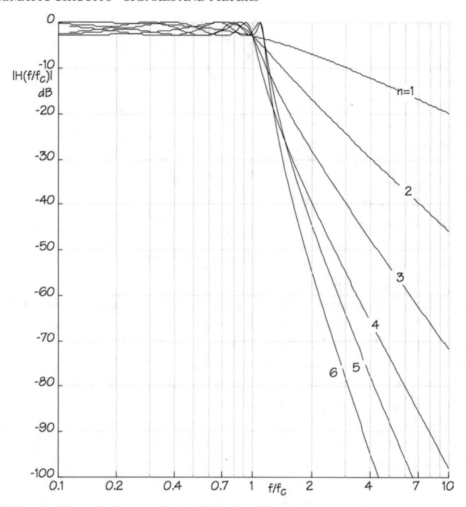

FIGURE 11.34: Normalized curves for the Chebychev filter.

I tried a number of commercial component values and finally settled on

$$R = 3.9 \text{ k}\Omega, C = 0.0022 \text{ µF},$$
$$R_f = 2 \text{ k}\Omega, R_1 = 1.2 \text{ k}\Omega.$$

Now let's match up the first-order stage to Fig. 11.24:

$$\left[\frac{s}{2\pi 20 \times 10^3 \times 0.2980} + 1\right]$$

$$\downarrow$$

$$[RCs + 1],$$

$$\text{so } RC = 26.704 \times 10^{-6}.$$

n=1	$(s/2\pi f_c)+1$
2	$(s/(0.8409\times 2\pi f_c))^2 +0.7654(s/(0.8409\times 2\pi f_c))+1$
3	$\left[(s/(0.2980\times 2\pi f_c))+1\right]\left[(s/(0.9159\times 2\pi f_c))^2 +0.3254(s/(0.9159\times 2\pi f_c))+1\right]$
4	$\left[(s/(0.9502\times 2\pi f_c))^2 +0.1789(s/(0.9502\times 2\pi f_c))+1\right]\left[(s/(0.4425\times 2\pi f_c))^2 +0.9276(s/(0.4425\times 2\pi f_c))+1\right]$
5	$\left[(s/(0.1772\times 2\pi f_c))+1\right]\left[(s/(0.9674\times 2\pi f_c))^2 +0.1132(s/(0.9674\times 2\pi f_c))+1\right]$
	$\left[(s/(0.6139\times 2\pi f_c))^2 +0.4670(s/(0.6139\times 2\pi f_c))+1\right]$
6	$\left[(s/(0.9771\times 2\pi f_c))^2 +0.0781(s/(0.9771\times 2\pi f_c))+1\right]\left[(s/(0.7223\times 2\pi f_c))^2 +0.2886(s/(0.7223\times 2\pi f_c))+1\right]$
	$\left[(s/(0.2978\times 2\pi f_c))^2 +0.9562(s/(0.2978\times 2\pi f_c))+1\right]$

FIGURE 11.35: Chebychev denominators.

I found values of

$$R = 2.7\ k\Omega,\ \ C = 0.01\ \mu F.$$

I can use the op-amp of this first-order stage to get the overall gain correct. The gain of the second stage is

$$k = 1 + \frac{2}{1.2} = 2.667.$$

We need 12 dB, which is 3.981. Hence this first-order stage must provide

$$\frac{3.981}{2.667} = 1.493 = 1 + \frac{R_f}{R_1},$$
$$R_f = 0.493 R_1,$$
$$\text{close to } R_f = 1\ k\Omega,\ R_1 = 2\ k\Omega.$$

That gives us the final design. Fig. 11.36 shows the final circuit and Fig. 11.37 is a plot of its frequency response. Note in the response curve the wiggling in the passband. But that wiggling stays between 12 dB and 9 dB, so it is within a 3-dB range. The cutoff frequency and the rolloff to 45 dB below the passband both are fine.

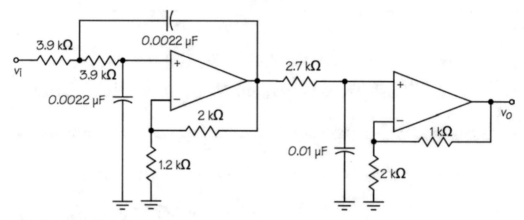

FIGURE 11.36: Chebychev design.

So which design is better? If "better" means fewer op-amps, two for Chebychev is better than three for Butterworth, which is *much* better than ten for cascaded first order.

There is one minor glitch in Chebychev design that requires some attention. If the order n is even, the gain will be too large by the square root of 2, which is 3 dB. Hence the final adjustment of overall gain must take this into account.

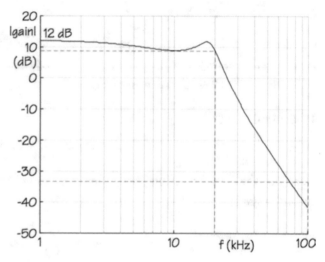

FIGURE 11.37: Chebychev design.

11.6 DESIGN EXAMPLE

A ham radio operator decides that she wants to build her own radio teletype system. She does, and the results seem pretty good. Her system is a frequency-shift keyed (FSK) signal that runs at the AMTOR (AMateur Teleprinting Over Radio) rate of 133 words per minute.

She gets on the air with her signal and "talks" to a friend across the country. While they are talking, another ham breaks in to tell her that she is splattering a signal several kilohertz above her communications frequency.

Improper operation such as this can earn a ham a citation from the Federal Communications Commission, so she decides that she better get off the air and see what the problem is.

She does some testing with a local ham friend and finds that she has a significant third harmonic of the teletype signal. It's probably caused by the fact that digital pulse circuits deal with…well, pulses, of course, and pulses are rich in harmonics.

Now comes her dilemma. Does she try to redo the digital circuitry to remove this problem? Or would it be easier to design a filter to remove, or at least reduce, the spurious signal? Since she's here in Chapter 11 at the moment, the filter seems like an obvious move.

But what to filter? She'll need to do a little reading, and the *ARRL Handbook* happens to be handy. From this she learns that 133 words per minute is 100 baud. Since this is two-level signaling, 100 baud is 100 bits per second. But what does this say about frequencies?

The standards for AMTOR are to use a frequency shift of 170 Hz, switching back and forth between 2125 and 2295 Hz. A simple formula from *The ARRL Handbook* gives the bandwidth that her filter must pass:

$$BW = B + DK,$$

where *BW* is the bandwidth that the signal will occupy, *B* is the signaling rate in baud, *D* is the peak frequency deviation in hertz, and *K* is an adjustment factor, normally about 1.2 for FSK operation.

Well, she knows all the numbers for this, so the bandwidth of her signal should be

$$BW = 100 + 170 \times 1.2 = 304 \, \text{Hz}.$$

That means that the signal is 304 Hz wide, somewhat wider than the 170-Hz deviation implies. The 170-Hz deviation is centered at 2210 Hz, half way between 2125 and 2295 Hz. The 304-Hz bandwidth is centered in the same place, so the signal occupies bandwidth from 2058 to 2362 Hz.

If that's the case, the third harmonic problem must be in the band between 6174 and 7086 Hz, three times the position of the fundamental. Great! Now she knows what the filter must cut off. But how large is the signal and how much must it be attenuated?

Her friend happens to be a student at the World's Best Engineering…, so he suggests taking the system into the lab and using a spectrum analyzer to figure out how big the problem is. After all sorts of problems, not the least of which revolves around the fact that this is a spectrum in the audio frequency range, they succeed in getting an idea of what the output of her AMTOR system looks like. This output is sketched in Fig. 11.38.

The third harmonics occupy a band just where she suspected, but they are larger than imagined. The peak of that band is only 6 dB (a factor of 2) below the fundamental, but the higher frequencies roll off rapidly.

Our ham decides, on the basis of her knowledge of FCC rules, that she needs to suppress the spurious harmonics 40 dB below the fundamental. That means an additional 34 dB. (The harmonic is already 6 dB down.)

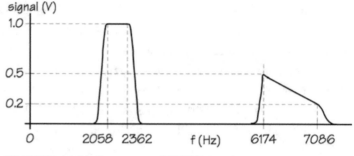

FIGURE 11.38: Spectrum of RTTY signal.

Now she starts a standard by-the-book filter design. The worst of the harmonics is at 6174 Hz. The upper end of the fundamental's band is 2362, and she doesn't want to alter that signal. The corner f_c is to be at 2362 Hz. The filter is to be down 34 dB at 6174 Hz. Normalizing this to f_c gives

$$\frac{f}{f_c} = \frac{6174}{2362} = 2.61.$$

She gets out her Butterworth and Chebychev normalized curves (Figs. 11.29 and 11.34) and finds that a fifth-order Butterworth or a third-order Chebychev filter will do the job.

Then she gets to thinking that Chebychev has a 3-dB ripple in the passband; for some reason she decides she'd rather not have that. She knows that there are 1-dB designs available, but she doesn't have any data on these. Moreover, while the Chebychev filter will require just two op-amps, the Butterworth filter will use just one more.

On top of all that, she notes that she can go to a sixth-order Butterworth filter with the same number of op-amps. That will allow her to position f_c a little above 2362 Hz so that the 3-dB corner won't alter her signal.

Her decision is to design a sixth-order Butterworth filter with 2362 Hz positioned at the 0.9 point on the f/f_c curves. That leaves some margin for error, both in the desired passband and at the frequency of the troublesome harmonic. So the design cutoff frequency f_c will be

$$f_c = 2362/0.9 = 2624 \text{ Hz}.$$

The Butterworth denominators are given in Fig. 11.30, and she chooses the one for $n = 6$. Each of these has the same second-order term:

$$\left(\frac{s}{2\pi f_c}\right)^2 = \left(\frac{s}{2\pi 2624}\right)^2 \approx \left(\frac{s}{16500}\right)^2.$$

Next, she goes to the standard second-order low-pass circuit of Fig. 11.22 and notes that the second-order term is

$$(RCs)^2 = \left(\frac{s}{16500}\right)^2,$$

$$RC = \frac{1}{16500}.$$

Trial-and-error using standard capacitor and resistor values yields

$$R = 6.2 \text{ k}\Omega, C = 0.01 \,\mu\text{F}.$$

Now to handle the first-order terms. Each of the first-order terms has a different coefficient in the sixth-order Butterworth denominator. Each of these must match up with the $(3 - k)$ term of the second-order low-pass filter she's chosen. The k here is the gain of the noninverter in the standard circuit that she's chosen. The calculations will lead to R_f and R_1 for each stage:

$$3 - k_1 = 0.5176, \ k_1 - 1 = 1.4824, \ R_{f1} = 1.4824 R_{11};$$
$$3 - k_2 = 1.414, \ k_2 - 1 = 0.586, \ R_{f2} = 0.586 R_{12};$$
$$3 - k_3 = 1.932, \ k_3 - 1 = 0.068, \ R_{f3} = 0.068 R_{13}.$$

After trying all sorts of resistor values, she discovers that fixing R_1 at 30 kΩ for all three stages yields decent results:

$$R_1 = 30 \ k\Omega,$$
$$R_{f1} = 42 \ k\Omega, R_{f2} = 18 \ k\Omega, R_{f3} = 2 \ k\Omega.$$

That completes the first design, but does it work? Since she has access to Maple, she gives Maple the overall $H(s)$ of her design and asks for a semilog plot. Fig. 11.39 shows the result.

The filter seems to have its 3-dB point a little close to the upper frequency in the passband. More-

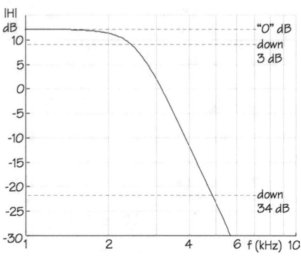

FIGURE 11.39: Initial design's response.

over, there's a lot of "space" to the right, because the filter is down 34 dB at about 5000 Hz instead of 6174 Hz.

Shifting this filter is easy. Our ham decides to move the 34-dB point from about 5000 to about 6000 Hz, a factor of 6/5. Since RC in the filter design is proportional to $1/f$, *decreasing* R by 5/6 should move the whole filter characteristic up by 6/5:

$$5/6\,R = 5/6 \times 6.2 = 5.17 \rightarrow 5.1\,\text{k}\Omega.$$

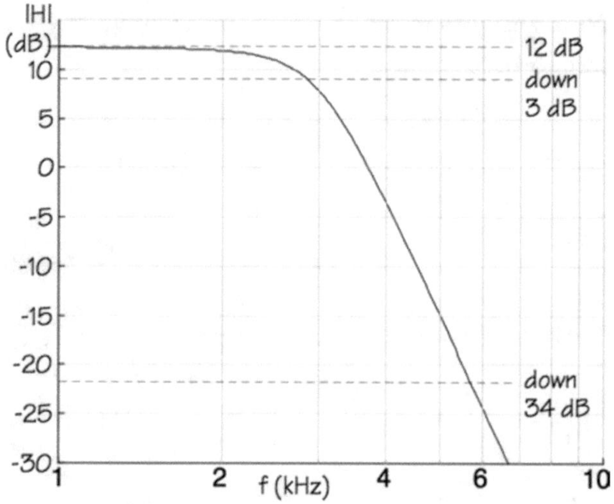

FIGURE 11.40: Final result.

With just this change she replots the characteristic of her filter (Fig. 11.40) and finds that everything is in the proper place.

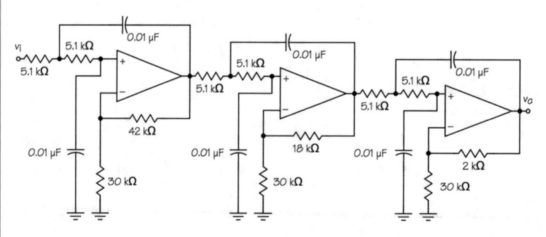

FIGURE 11.41: Final design: sixth-order Butterworth filter.

So she builds the circuit shown in Fig. 11.41 and her AMTOR operation works fine. Nobody reports that she has signal splatter where she doesn't belong. And everyone lives happily every after. (Especially after she publishes in *QST*.)

11.7 SUMMARY

Where have we been in this chapter? Mostly through low-pass filters, and those "by the book." There are several things to learn from all of this:

- There are circuits that make good building blocks for filters and the like. It makes little sense to take time to reinvent a circuit if a good one already exists.
- Building blocks are generally given on a unit basis, where things basically have values of 1. When we use these blocks we scale the values to fit our needs.
- Cascaded first-order design is easy to do but uses lots of op-amps. We might choose this design when we want to keep our circuit very basic.
- Cascaded second-order designs generally use fewer op-amps to achieve the same result as the first-order designs.
- Butterworth and Chebychev are particular forms of second-order designs where the individual stages vary in clever ways.
- Butterworth filters have their poles on a circle. They are maximally flat and have better cutoff characteristics than the first order designs. They are synchronously tuned, which means that all stages have the same RC value.
- Chebychev filters improve on Butterworth filters by placing the poles in an elliptical pattern. They achieve even better cutoff characteristics but have ripple in the passband. The stages are stagger tuned so that each has a different RC value.
- "By the book" design is a useful and quick way of getting results. It allows us to use building blocks and standard designs, and that saves time.

Where next? Remember several chapters ago that we worked with the Fourier series? This was a way to represent as sums of sinusoidal functions that are periodic. But what do we do with signals that are not so neatly organized? How do we represent them?

The Fourier transform in the next chapter will deal with nonperiodic signals. This transform enables us to get the spectrum of such a signal, and through that, to determine what happens as the signal passes through communications channels.

CHAPTER 12

Fourier Transforms:
Oh, no! Not Another One

"I've got transforms coming out my ears and you're going to do another one? First it's the sinusoidal steady state transformed into the phasor domain. Then we had Laplace and, yes, I know, it moves us from the *t* domain to the *s* domain. Recently it was the Fourier series to describe periodic signals as phasors. One of my friends even had an instructor tell the class about delta-wye transforms. Or was it transformers? I dunno. But now Fourier… this is too much!"

Well, our somewhat annoyed student perhaps has a point. We have done three major transformations:

- systems operating in the steady state with sinusoidal inputs can be efficiently represented by phasors, thereby transforming the time domain into the phasor domain;
- systems whose activities start at $t = 0$ can be neatly represented using the Laplace transform, which transforms the time domain into the frequency (s) domain; and
- periodic signals can be represented by the Fourier series, again transforming the time domain into the phasor domain.

Now we add the Fourier *transform* that allows us to transform the time domain into the frequency domain again. This time, instead of the complex s, we will be restricting our attention to just $j\omega$ (or really just jf). But where the Laplace transform requires us to start at $t = 0$, the Fourier transform does not. And where the Fourier series requires us to work with periodic signals, the Fourier transform does not.

What does this do for us? Suppose we have a single square pulse to pass through a communications channel. What does this bandlimited channel do to such a signal? In these days of digital systems, that's an important question.

12.1 FOURIER TRANSFORM MATH

Please think back over a couple of things that we have done in previous chapters:

- the idea of a *spectrum* that describes in the frequency domain what kinds of frequencies are present in a signal, including both magnitude and phase at different frequencies;

- the *discrete spectrum* produced by the Fourier series, again including both magnitude and phase, which show that the signal has energy only at certain frequencies.

The Fourier *transform* will generate for us spectra for nonperiodic signals. The results will be spectra that are *continuous*. In other words, while certain frequency components may be more prominent than others, the spectra will be continuous functions that will span all frequencies from $-\infty$ to $+\infty$.

Now we can bury ourselves in mathematics—if we want to! But I don't want to. There are several ways to derive the integral transformations that are the Fourier transform. I'll leave those to others because I am quite willing to believe Mr. Fourier.

One bit of warning. As I have done in a number of places in this text, I am going to use *f* as the independent frequency variable. In other words, I am writing these functions in *hertz* rather than radians per second. That's done in texts like this one, but rarely in math books.

What will this difference mean? If you happen to pick up a math book that uses ω, you'll find that 2π keeps showing up. I'll point out one important place shortly.

If you are really mathematically inclined (i.e., if you have no practical bone in your body), the following pair of integrals is all you will ever need to know. They are the integral transformations that define the Fourier transform.

But...as you'll see as we go along, our approach to the Fourier transform is going to be a lot more practical. Our goal is to *understand* what happens in the frequency domain when we deal with certain time functions, and especially when we modify them. Our goal is also to *understand* just the opposite, namely, what happens in the time domain when we deal with certain frequency functions, and especially when we modify them.

So here are the integrals of the Fourier transform pair, keeping in mind that we won't actually carry them out very often:

$$X(f) = \int\limits_{-\infty}^{+\infty} x(t)e^{-j2\pi ft}\,dt,$$

$$x(t) = \int\limits_{-\infty}^{+\infty} X(f)e^{+j2\pi ft}\,df.$$

Let's take a tour of these, noting a few of the landmarks as we go:

- *X(f)* is the Fourier transform of *x(t)*, and *x(t)* is the inverse transform of *X(f)*;
- the pair is unique, meaning that one of the integrals transforms one function into another, while the other integral brings the first function back;
- the integrations are over all time or all frequency, so we are not dealing, as in the Laplace transform, with signals that must start at $t = 0$;

- neither integral is symmetric about the origin—the first has a minus in the exponent, the second, a plus;
- the usual form of these integrals is to have $j\omega t$ in the exponent, but I have chosen to use $j2\pi f t$; and
- because of this, the usual $1/2\pi$ that would appear in front of the second integral in math books is missing. This change in the 2π factor will show up in other places as well.

As with anything that we do in math (so it seems), there are restrictions. Here, the restriction is very simple:

$$\int_{-\infty}^{+\infty} |x(t)| \, dt < \infty.$$

This says that if you can't integrate the magnitude of the time-domain function over all time and get a finite value, you can't have a Fourier transform of that function.

Except that this is a bit of a lie. We will get away with violating the condition for a number of functions. Two in particular that have Fourier transforms but don't meet the condition are $x(t) = 1$ and $x(t) = \cos 2\pi f t$.

An example seems like a good way to finish off this section. The signal shown in Fig. 12.1 is ten periods of a cosine, centered at $t = 0$. Think of it as a little burst of a tone whose frequency is 5 kHz. It is on for only 2 ms.

This function is described mathematically by

$$x(t) = \begin{cases} 4\cos(2\pi\, 5000t) & -1 < t < 1\,\text{ms}, \\ 0 & \text{elsewhere.} \end{cases}$$

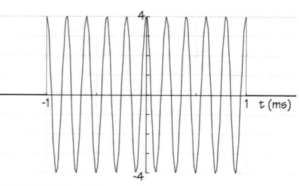

FIGURE 12.1: Cosine pulse.

I did the integration to get

$$X(f) = \int_{-0.001}^{0.001} 4\cos(2\pi\, 5000t)\, e^{-j2\pi f t}\, dt$$

$$= \frac{4f \sin(\pi f/500)}{\pi\left(f^2 - 25000000\right)}$$

$$= \frac{4}{\pi}\, \frac{f \sin(0.002\pi f)}{f^2 - 5000^2}\, \text{V/Hz}.$$

Oh, my! Now what? There are a number of things to look at in this function, and all of them tell something about the frequency spectrum of this pulsed cosine:

- The amplitude of the pulse (4) is sitting out in front. But this might mislead us in figuring out the units of this function. "4" is in volts, f is in Hz, and f in the denominator is in Hz². (The sine is unitless.) So the units are *volts per hertz*.
- It's pretty clear that the denominator will go to zero when $f = \pm 5000$ Hz. So this seems to imply that the function $X(f)$ will be infinite. But note that $\sin(0.002\pi f)$ is also zero when $f = 5000$. So we have a case of $0/0$. L'Hopital told us how to do this, and doing it yields $X(5000) = 0.004 = 4$ mV/Hz.

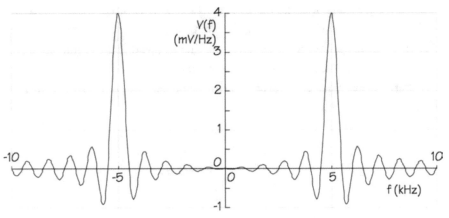

FIGURE 12.2: Fourier transform of a cosine pulse.

So? So even with that information we still don't have much of a feeling for this function in the frequency domain. Plotting it does a lot more for us (Fig. 12.2).

Note that the horizontal axis is in kilohertz. Those peaks? These are the places where $f = \pm 5$ kHz. The wiggles on either side of each peak show that the voltage falls off as we go away from the peaks.

But what is this spectrum? Let's look at it from an energy standpoint. Recall that we use the 1-Ω energy to talk about energy in a spectrum. Here we have voltage, and not really voltage but rather *voltage density* (volts per unit frequency). If we simply squared the function, we'd have *energy density*.

We don't need to do the actual squaring to understand, though, that the energy in this signal is concentrated near 5 kHz. The energy is concentrated near the frequency of the cosine that we used in our pulse.

But there is energy at other frequencies, too. This comes from the fact that the pulse is cut off very sharply. More sharply, in fact, than we can ever achieve in a real system. (Recall that we can never create a step, only an approximation of it to whatever closeness we can afford.)

The shape that we have gotten in this example is of the form (sin x)/x. This is going to show up many times as we continue with Fourier transforms. And you might recall that in Chapter 10 we also got one, with a note that said more are coming. They are here.

12.2   MEXICAN HAT

Our goal in this chapter is to make use of the Fourier transform to transform nonperiodic signals from the time domain to the frequency domain. To say it another way, our goal is to use the Fourier transform to get frequency spectra of signals.

If that's the goal, then a subgoal is going to be to avoid doing all sorts of mathematics just to get to the spectrum and back again. We'd like to get to the point where we can, with some straightforward thinking, obtain the spectrum. After all, we'd like to understand what is going on in our system, not spend lots of time doing math.

As a start toward this, let's consider a simple pulse. The pulse shown in Fig. 12.3 has a height of 1 V and lasts from $t = -1$ to $+1$ s:

$$v(t) = \begin{cases} 1 & -1 < t < +1 \\ 0 & \text{elsewhere.} \end{cases}$$

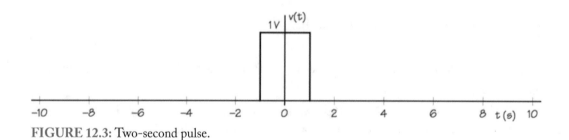

FIGURE 12.3: Two-second pulse.

Transforming this mathematically is easy because the integral is rather simple:

$$V(f) = \int_{-1}^{+1} 1 e^{-j2\pi ft} \, dt$$

$$= 2 \frac{\sin 2\pi f}{2\pi f} \text{ V/Hz.}$$

Aha! There's that (sin x)/x again. I've cleaned up the result a little so that the factor $2\pi f$ appears intact in both the numerator and the denominator.

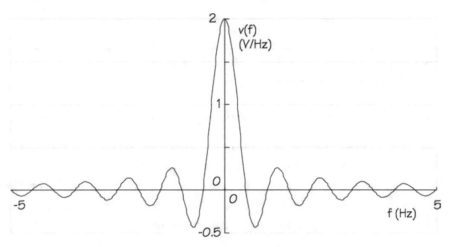

FIGURE 12.4: Fourier transform of a two-second pulse.

It's rather clear, I think, that something must happen at $f = 0$ because the denominator will be zero. But so will the numerator. The result is a classic (sin x)/x display, as shown in Fig. 12.4.

Note three things about this:

- The energy is concentrated at $f = 0$, which should be no surprise because the pulse is basically DC while it is on.
- There is energy at frequencies away from $f = 0$, caused by the abrupt changes in the pulse at $t = -1$ and $t = +1$ s.
- The quantity outside the (sin x)/x term (namely, 2) is the peak of the curve. The units are volts per hertz.

Also note that the width of the pulse, which is 2 s, appears explicitly in the $2\pi f$ factor.

Now let's extend the pulse. I'll keep it at the same height but start it at -10 s and continue until $+10$ s. Fig. 12.5 shows this. Keep in mind that I now have a pulse that contains ten times as much energy as did the first one.

If we compute the Fourier transform we get

$$v(t) = \begin{cases} 1 & -10 < t < +10 \\ 0 & \text{elsewhere,} \end{cases}$$

$$V(f) = 20 \frac{\sin 20\pi f}{20\pi f} \text{ V/Hz.}$$

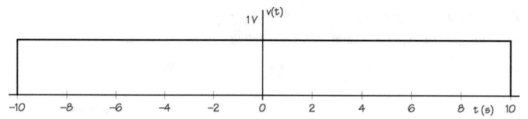

FIGURE 12.5: Fourier transform of two-second pulse.

Again, there's the (sin x)/x, which I have cleaned up as before. There are two things to note:

- The number in front is 20, or ten times what we got for the first pulse. This reflects the fact that the pulse contains ten times as much energy.
- The factor $20\pi f$ inside the sine term means that the sine term wiggles *ten times* as fast as the previous one wiggled. The plot of Fig. 12.6 shows this clearly. (I will keep all the plots in this section on the same-sized horizontal axes.)

Well, OK, now what? Let's do this one more time (before we get too bored) by trying a pulse (Fig. 12.7) that extends only from t = −0.1 to t = +0.1 s. This pulse must contain only one-tenth the amount of energy as the first one.

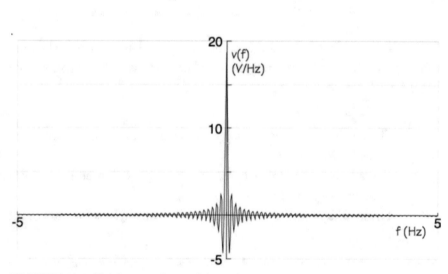

FIGURE 12.6: Fourier transform of the pulse ten times as wide.

$$v(t) = \begin{cases} 1 & -0.1 < t < +0.1 \\ 0 & elsewhere, \end{cases}$$

$$V(f) = 0.2\frac{\sin 0.2\pi f}{0.2\pi f} \text{ V/Hz}.$$

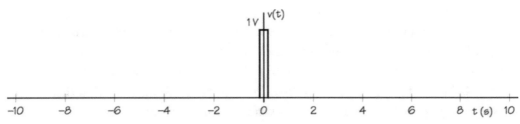

FIGURE 12.7: One-tenth as wide.

Hmmm, the number in front is now also one-tenth as large. And the pulse duration still appears in the $(\sin x)/x$ term explicitly (0.2). But because this number is now smaller, the wiggles should take even longer than they did in the first case.

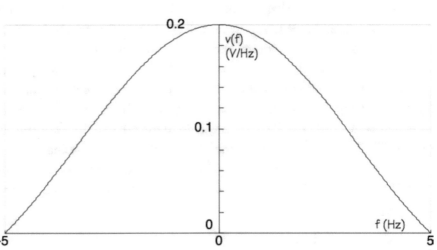

FIGURE 12.8: Fourier transform of the pulse one-tenth as wide.

Fig. 12.8 shows this.

Since I have kept the same horizontal scale for all three, the wiggles don't even show here.

Let's take the big step and draw some conclusions, rather important ones:

- As the pulse gets wider, the spectrum gets narrower.
- As the pulse gets narrower, the spectrum gets wider.
- As the 1-Ω energy in the pulse increases, so does the amplitude of the spectrum.

We will use these observations to our advantage as we continue working with Fourier spectra.

So that finishes another example, unless you are wondering where the title of this section came from. Try plotting $(\sin x)/x$ in three dimensions. Plot the function and see what you get in 3D. (Replace x by $\sqrt{x^2 + y^2}$ and let both x and y extend over the range of ± 10.)

12.3   USEFUL PROPERTIES

Now let's do something that may seem out of character. I've pretty much made it a habit throughout this text to avoid mathematical details and formulas. This section will look to you, perhaps, like I've lost that focus. It's going to be full of details about the Fourier transform.

But, no, I haven't gone off the deep end or been abducted by mathematical aliens! This section on some properties of the Fourier transform has a higher purpose.

Recall that earlier I said we'd like to be able to do much of our transform work without doing the integrations. More than that, we'd like to be able to look at functions in the time domain and know about what kind of spectra they'll have. We'd like to be able to consider what a change to a signal in the time domain will do to its spectrum. And we'd like to be able to do these same things from the spectra back to the time domain.

Consider a simple example. Suppose I have a signal whose spectrum I know. I am working with that spectrum and a certain bandlimited communications channel (*all* channels are bandlimited). Suppose that I must further limit the bandwidth of the signal. It would be nice to know approximately what this will do to the original time-domain signal without integrating. (In fact, sometimes the alterations are such that integrating would be very messy.)

So the purpose of going through some of the properties of the Fourier transform is to provide tools for your toolbox that will increase your ability to work with signals and spectra efficiently.

12.3.1 Even and Odd

Let's take the Fourier transform and have Mr. Euler work on it a bit:

$$X(f) = \int_{-\infty}^{\infty} x(t)e^{-j2\pi ft}\,dt$$

$$= \int_{-\infty}^{\infty} x(t)\cos(2\pi ft)\,dt - j\int_{-\infty}^{\infty} x(t)\sin(2\pi ft)\,dt.$$

It is fairly common to use A and B to label these:

$$A(f) = \int_{-\infty}^{\infty} x(t)\cos(2\pi ft)\,dt,$$

$$B(f) = -\int_{-\infty}^{\infty} x(t)\sin(2\pi ft)\,dt,$$

$$\text{so } X(f) = A(f) + jB(f).$$

Since A and B combine as a complex number, we can also conclude that

$$|X(f)| = \sqrt{A^2(f) + B^2(f)},$$

$$\angle X(f) = \tan^{-1}\frac{B(f)}{A(f)}.$$

So $X(f)$ has both magnitude and phase just like any other spectrum. Now several observations are in order:

- $A(f)$ is an even (cosine) function, so $A(f) = A(-f)$;
- $B(f)$ is an odd (sine) function, so $B(f) = -B(-f)$;
- $|X(f)|$ is even because of the squares;
- $\angle X(f)$ is odd because the arctangent of an angle is the negative of the arctangent of the negative of that angle;
- $X(-f) = X^*(f)$, the complex conjugate;
- if $x(t)$ is an even function, $B(f) = 0$ because the integral over all time of an even function times an odd function is zero; and
- if $x(t)$ is an odd function, $A(f) = 0$ for the reason just stated.

I think you'll want to remember most of those, or have them where you can use them easily.

12.3.2 Linearity

The Fourier transform is linear, which means two things. First, we can add transforms of added functions. Second we can multiply by a constant.

$$\text{If } X_1(f) = \int_{-\infty}^{\infty} x_1(t)e^{-j2\pi ft}\,dt,$$

$$\text{and } X_2(f) = \int_{-\infty}^{\infty} x_2(t)e^{-j2\pi ft}\,dt,$$

$$\text{then } X_3(f) = X_1(f) + X_2(f)$$

$$= \int_{-\infty}^{\infty} [x_1(t) + x_2(t)]e^{-j2\pi ft}\,dt;$$

$$\text{also } X_4(f) = \int_{-\infty}^{\infty} ax_1(t)e^{-j2\pi ft}\,dt$$

$$= a\int_{-\infty}^{\infty} x_1(t)e^{-j2\pi ft}\,dt$$

$$= aX_1(f).$$

To say this all another way, if we add two signals, we add their transforms. (That's superposition.) If we double the voltage of a signal, we double its transform.

12.3.3 Scaling, Flipping, and Translating

Suppose we have a simple transform:

$$X_1(f) = \int_{-\infty}^{\infty} x(t)e^{-j2\pi ft}\, dt.$$

Suppose further that we *scale* the time axis by replacing all values of t by t multiplied by the constant a. Let's do that in the integral and massage things a little:

$$X_2(f) = \int_{-\infty}^{\infty} x(at)e^{-j2\pi fat}\, d(at)$$

$$= a\int_{-\infty}^{\infty} x(at)e^{-j2\pi fat}\, dt.$$

But now if we replace f by f'/a to get rid of the a in the exponent, we get

$$X_2(f) = a\int_{-\infty}^{\infty} x(at)e^{-j2\pi f't}\, dt.$$

The integral on the right-hand side is a times the transform of the function $x(at)$ using f' as the frequency. We can finish this by recording that fact, then rearranging things, and finally getting f' out of it:

$$X_2(f) = a\mathscr{F}[x(at)],$$

$$\mathscr{F}[x(at)] = \frac{1}{a}X_2(f') = \frac{1}{a}X_2\left(\frac{f}{a}\right).$$

To say that in English, if I divide time by a constant, I multiply the frequency axis and the magnitude of the spectrum by that constant.

But even that isn't quite clear unless you look at which way things are going. Let's presume that the constant is 10. That means that a specific point on the time axis that was labeled "1 s" would now be labeled "0.1 s." To say this another way, whatever was happening in the signal at $t = 1$ is now happening at $t = 0.1$. That's much *faster*.

Suppose this particular signal had a feature in the spectrum at $f = 50$ Hz. The new frequency is the original frequency multiplied by the time-scaling factor, so the same feature now happens at 500 Hz. Moreover, the feature will be only 1/10 as large.

Although I may be trying to slaughter a point, this says that if I *slow down* a signal, its bandwidth gets *smaller*. That certainly makes sense! If I slow down a signal, it doesn't wiggle as fast, so it doesn't contain as many higher frequencies.

Whew! And by the way, we can work this scaling business from frequency back to time. I'll bet you'll let me do it without proof, too, and without lots of words:

$$\text{If } \mathcal{F}\left[x(t)\right] = X(f),$$

$$\text{then } X(af) = a\mathcal{F}\left[x\left(\frac{t}{a}\right)\right].$$

Now that we can change the size of the time axis, how about flipping the time axis over? What will happen to our transform when we replace t by $-t$?

$$\text{If } \mathcal{F}\left[x(t)\right] = X(f),$$

$$\mathcal{F}\left[x(-t)\right] = ?$$

We start by writing the standard Fourier transformation and then replacing t by $-t$:

$$\mathcal{F}\left[x(t)\right] = \int_{-\infty}^{\infty} x(t)e^{-j2\pi ft}\,dt,$$

$$\mathcal{F}\left[x(-t)\right] = \int_{-\infty}^{\infty} x(-t)e^{-j2\pi ft}\,dt.$$

We'll get somewhere with this if we replace $-t$ by t', keeping in mind that we must change t everywhere it appears in the integral, including the limits, which we exchange:

$$\mathcal{F}\left[x(-t)\right] = \int_{\infty}^{-\infty} x(t')e^{-j2\pi f(-t')}\,d(-t')$$

$$= -\int_{-\infty}^{\infty} x(t')e^{-j2\pi(-f)t'}\,d(t')$$

$$= X(-f).$$

(If you weren't paying attention you didn't notice the neat rearrangement of the signs in the exponent!)

So the answer to our question is fairly simple. If we flip the time axis over, we flip the frequency axis over:

$$\text{If } \mathcal{F}[x(t)] = X(f),$$
$$\mathcal{F}[x(-t)] = X(-f).$$

And that's only a reversal of the phase angle; the magnitude will remain the same.

Now that we have scaled the time axis and also flipped it over, we need to consider what happens if we move the origin on the time axis. A strange thing. Let's guess a little. Moving the origin moves the signal in time. But since we integrate over all time, the frequency content of the signal certainly is not being changed. So we should suspect that the transform should still look the same.

Almost. Moving the origin moves the starting point. The starting point in the time domain defines, in the frequency domain, the phase reference. So we should expect the phase angle to change, or at least shift in some way.

Let's do this by replacing t by $t - a$ (dt is still dt because a is a constant):

$$\mathcal{F}[x(t-a)] = \int_{-\infty}^{\infty} x(t-a)e^{-j2\pi f(t-a)}\,dt.$$

Now replace $t - a$ by t' (dt becaomes dt'):

$$\mathcal{F}[x(t-a)] = \int_{-\infty}^{\infty} x(t')e^{-j2\pi f(t'+a)}\,dt'$$

$$= e^{-j2\pi fa} \int_{-\infty}^{\infty} x(t')e^{-j2\pi ft'}\,dt'.$$

But whether we integrate over all t or all t' makes no difference, so

$$\mathcal{F}[x(t-a)] = e^{-j2\pi fa}\mathcal{F}[x(t)].$$

Aha! There's the transform of the original (untranslated) time function, but preceded by...well, by what? That thing is a phase angle, which we could write using the angle notation: $\underline{/-2\pi fa}$. That means the phase spectrum is shifted by $-2\pi fa$.

That's correct if we think about it. If we shift the time axes by, say, 5 s, an event that occurred at $t = 15$ s now will occur at $t = 10$ s. That's a *smaller* delay and hence a *smaller* phase angle.

It is equally easy to show translation from the spectrum to the time domain:

$$\text{If } x(t) = \mathscr{F}^{-1}\big[X(f)\big],$$

$$\text{then } e^{j2\pi f_a t} x(t) = \mathscr{F}^{-1}\big[X(f - f_a)\big].$$

12.3.4 Impulse, Constant, Signum, & Step

Several rather simple functions are important, not just because we can find Fourier transforms for them but also because they will come up when we work with common signals. I'll start with the impulse in the time domain.

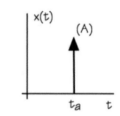

FIGURE 12.9: Impulse $\delta(t-t_a)$.

Fig. 12.9 shows an impulse in the time domain. It "happens" at $t = t_a$ and its area is A. The integration is easy to set up and easy to carry out. Easy, that is, if we remember that the impulse has a value only at exactly $t = t_a$. So the integral reduces to evaluating the rest of the stuff under the integral sign at just $t = t_a$:

$$X(f) = \int_{-\infty}^{\infty} A\delta\left(t - t_a\right) e^{-j2\pi ft}\, dt$$

$$= Ae^{-j2\pi ft_a},$$

which has a magnitude of A for all frequencies f, along with a phase angle $-2\pi ft_a$ that gets more and more negative as the frequency increases.

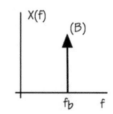

FIGURE 12.10: Impulse $\delta(f-f_b)$.

How about going the other way? Fig. 12.10 shows an impulse in the frequency domain. The integration is about the same as the one we just did:

$$x(t) = \int_{-\infty}^{\infty} B\delta\left(f - f_b\right) e^{+j2\pi ft}\, df$$

$$= Be^{+j2\pi f_b t}$$

$$= B\cos\left(2\pi f_b t\right) + jB\sin\left(2\pi f_b t\right).$$

We'll see this one again.

Now let's try a constant A as shown in Fig. 12.11. The math seems like it is going to be easy:

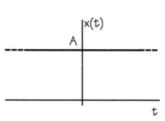

FIGURE 12.11: Constant A.

$$x(t) = A,$$

$$X(f) = \int_{-\infty}^{\infty} A e^{-j2\pi ft} \, dt$$

$$= \frac{1}{-j2\pi f} A e^{-j2\pi ft} \Big|_{-\infty}^{\infty}$$

$$= ?$$

Hmmm, that didn't work because we can't find a value for $e^{-j2\pi f\infty}$. Moreover, this should not have worked, because there's a restriction on the Fourier transformation:

$$\int_{-\infty}^{\infty} |x(t)| \, dt < \infty.$$

Perhaps we can get this by using our heads instead of blind mathematics? This function A is a constant for all time. That sure sounds like DC. And DC is zero frequency. So the spectrum must be concentrated at $f = 0$.

What should we draw at $f = 0$ for our spectrum? A dot? No, not likely, because we still have to be able to integrate this dot using the inverse transform and get back our DC signal.

How about a vertical line like those we drew several chapters ago? Nope, that won't work for the same reason as the dot.

Whatever we do, it must integrate so that we get A back again when we return to the time domain. I'll guess that the function is an impulse that should have the area A as shown in Fig. 12.12 and see what happens.

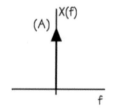

FIGURE 12.12: Impulse $\delta(f)$.

Writing the integral and then integrating is easy again because the integral "happens" only at the impulse, which is $f = 0$:

$$x(t) = \int_{-\infty}^{\infty} A\delta(f) e^{j2\pi ft} \, df$$

$$= A e^{j2\pi(0)t} = A,$$

so $\mathcal{F}[A] = A\delta(f)$.

Success!

The next function we need is signum, a rather strange function whose value is +1 if its argument is positive and −1 if

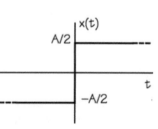

FIGURE 12.13: Signum(t).

its argument is negative. Fig. 12.13 shows the function $(A/2)$ signum(t). Note that it resembles a symmetric step.

This function violates the restriction on the Fourier transformation, too, so we get to the result through an artifice. Let's approximate the signum function with two exponentials:

$$x(t) = \begin{cases} \dfrac{A}{2} e^{-at} & t > 0 \\[2mm] -\dfrac{A}{2} e^{at} & t < 0 \end{cases}.$$

This is an odd function so I can use the odd property of the transform:

$$\mathcal{F}\left[\frac{A}{2} signum(t)\right] = 2\int_0^\infty \frac{A}{2} e^{-at} e^{-j2\pi ft}\, dt$$

$$= A\int_0^\infty e^{-(a+j2\pi f)t}\, dt$$

$$= \frac{-A}{a+j2\pi f} e^{-(a+j2\pi f)t}\Bigg|_0^\infty$$

$$= \frac{A}{a+j2\pi f}.$$

To finish this, let the exponential approximation to the signum function approach signum. If we make a in the exponential approximation go to zero, the exponential looks like $(A/2)$ signum(t) in the limit. The limit applied to the transform yields

$$\lim_{a\to 0} \frac{A}{a+j2\pi f} = \frac{A}{j2\pi f}.$$

Fig. 12.14 below shows what this form looks like.

Finally in this section we need the step function, shown in Fig. 12.15. I'll do this the easy way by using the signum function we just developed and adding $A/2$ to it to boost it up to become a step of size A:

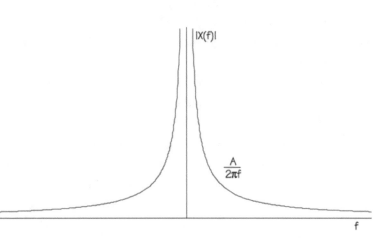

FIGURE 12.14: $X(f)$ for $(A/2)$signum(t).

$$\mathcal{F}\left[Au(t)\right]=\mathcal{F}\left[\frac{A}{2}signum(t)\right]+\mathcal{F}\left[\frac{A}{2}\right]$$

$$=\frac{A}{j2\pi f}+\frac{A}{2}\delta(f).$$

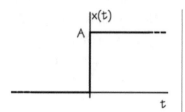

FIGURE 12.15: Step $u(t)$.

Fig. 12.16 shows the form of this transformation. This says that the step looks a lot like DC, but the sharp corners at $t = 0$ add lots of other frequencies to the spectrum.

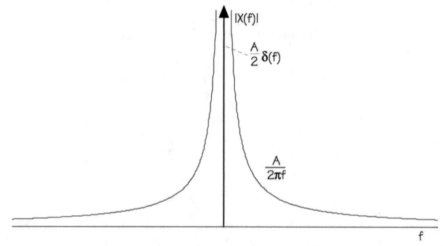

FIGURE 12.16: $X(f)$ for $Au(t)$.

12.3.5 Differentiation and Integration

Suppose we differentiate $x(t)$. What happens to the transform? We should suspect something simple because of what happened with the Laplace transform, where differentiation was a multiplication by s.

It is easier (for me, at least) to get at this by working with the inverse transform. If $X(f)$ is the Fourier transform of $x(t)$, then

$$x(t) = \mathcal{F}^{-1}\left[X(f)\right]$$

$$= \int_{-\infty}^{\infty} X(f)e^{+j2\pi ft}\,df.$$

Now differentiate $x(t)$ and hence the integral:

$$\frac{dx}{dt} = \frac{d}{dt}\left[\int_{-\infty}^{\infty} X(f)e^{+j2\pi ft}\,df\right].$$

But the differentiation and the integration can be interchanged because t appears only in the exponential. Doing that and then finishing the differentiation gives

$$\frac{dx}{dt} = \int_{-\infty}^{\infty} X(f)\frac{d}{dt}\left[e^{+j2\pi ft}\right]df$$

$$= \int_{-\infty}^{\infty} j2\pi f X(f)e^{+j2\pi ft}\,df$$

$$= \mathscr{F}^{-1}\left[j2\pi f X(f)\right].$$

That last line says that the inverse transform of $j2\pi f$ times a transform yields the derivative of the time function. That means that

$$\mathscr{F}\left[\frac{dx(t)}{dt}\right] = j2\pi f\,\mathscr{F}\left[x(t)\right].$$

Not unlike the Laplace transform, eh? The derivative in the time domain is multiplication by $j2\pi f$ in Fourier's frequency domain. (Recall, though, that the Laplace transform included the initial condition, something Fourier doesn't deal with.)

The integration property is likewise obvious, or sort of. We'd expect that integration in the time domain would be division by $j2\pi f$ in the frequency domain. It will turn out that way...almost.

I'll start with the transform of the time integral of a function $x(t)$:

$$\mathscr{F}\left[\int_{-\infty}^{t} x(\tau)\,d\tau\right] = \int_{-\infty}^{\infty}\left[\int_{-\infty}^{t} x(\tau)\,d\tau\right]e^{-j2\pi ft}\,dt.$$

(Note the use of the dummy variable of integration.)

This mess wouldn't be quite so much of a mess if both integrals had the same limits. I can do this by including in the time function a "backward step" that will "turn off" the function $x(\tau)$ at t. Then I can run the inside integral all the way to $+\infty$. The function is shown in Fig. 12.17. (Note that the horizontal axis is τ, not t.)

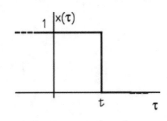

FIGURE 12.17: $u(-\tau+t)$.

Including the function in the integral gives us

$$\mathcal{F}\left[\int_{-\infty}^{t} x(\tau)d\tau\right] = \int_{-\infty}^{\infty}\left[\int_{-\infty}^{\infty} u(t-\tau)x(\tau)d\tau\right]e^{-j2\pi ft}\,dt.$$

But now I can do the integrals in either order, so I'll rearrange the terms a bit:

$$\wedge\left[\int_{-\infty}^{t} x(\tau)d\tau\right] = \int_{-\infty}^{\infty}\left[\int_{-\infty}^{\infty} u(t-\tau)e^{-j2\pi ft}\,dt\right]x(\tau)d\tau.$$

Now I recognize the thing in the brackets to be the Fourier transform of a step. The step is the unit step and hence has a size of 1. It is delayed by the amount τ. So I can replace the stuff in the brackets by

$$\left[\frac{1}{j2\pi f}+\frac{1}{2}\delta(f)\right]e^{-j2\pi f\tau}.$$

(The brackets enclose the transform of the step; the term outside the brackets is the delay by an amount τ.)

I'll do the replacement. Then I'll pull out of the remaining integral anything that is not a function of t. What is left is the transform of $x(t)$:

$$\mathcal{F}\left[\int_{-\infty}^{t} x(\tau)d\tau\right] = \int_{-\infty}^{\infty}\left[\left[\frac{1}{j2\pi f}+\frac{1}{2}\delta(f)\right]e^{-j2\pi f\tau}\right]x(\tau)d\tau$$

$$= \left[\frac{1}{j2\pi f}+\frac{1}{2}\delta(f)\right]\int_{-\infty}^{\infty} e^{-j2\pi ft}x(\tau)d\tau$$

$$= \left[\frac{1}{j2\pi f}+\frac{1}{2}\delta(f)\right]\mathcal{F}\left[x(t)\right].$$

This can be cleaned up a little by rearranging:

$$\mathcal{F}\left[\int_{-\infty}^{t} x(\tau)d\tau\right] = \frac{1}{j2\pi f}\mathcal{F}\left[x(t)\right]+\frac{1}{2}\delta(f)\mathcal{F}\left[x(t)\right]$$

$$= \frac{1}{j2\pi f}\mathcal{F}\left[x(t)\right]+\frac{1}{2}\delta(f)X(f).$$

Now I'll note that the last term has a value only at $f = 0$ because of the impulse at $f = 0$. So $X(f)$ can be evaluated rather easily:

$$X(f = 0) = \int_{-\infty}^{\infty} x(t)e^{-j2\pi(0)t}\,dt$$

$$= \int_{-\infty}^{\infty} x(t)\,dt.$$

This says that $X(0)$ is the DC value of $x(t)$. If there is no DC value, the term containing the impulse disappears, leaving us with just what we had guessed from our knowledge of Laplace.

So time integration in the time domain transforms into a division by $j2\pi f$ in Fourier's domain—if there is no DC value. If there's a DC value, we must add an impulse.

$$\mathcal{F}\left[\int_{-\infty}^{t} x(\tau)\,d\tau\right] = \frac{1}{j2\pi f}\mathcal{F}\left[x(t)\right] + \frac{1}{2}X(0)\delta(f).$$

12.3.6 Exponential

The double-sided exponential shown in Fig. 12.18 is something useful. We'll find its Fourier transform by doing only the positive half and then using the flipping property to get the whole story:

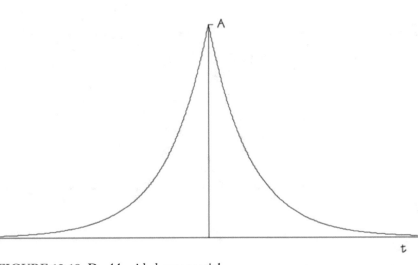

FIGURE 12.18: Double-sided exponential.

Here's the integration for the positive half:

$$x_+(t) = Ae^{-at}u(t),$$

$$X_+(f) = \int_{-\infty}^{\infty} Ae^{-at}u(t)e^{-j2\pi ft}\,dt$$

$$= \int_0^{\infty} Ae^{-(a+j2\pi f)}\,dt$$

$$= \frac{-A}{a+j2\pi f}e^{-(a+j2\pi f)t}\Big|_0^{\infty}$$

$$= \frac{A}{a+j2\pi f}.$$

If we use the flipping property that we've already derived, we get

$$x_-(t) = Ae^{+at}u(-t),$$

$$X_-(f) = \frac{A}{a-j2\pi f}.$$

Now use linearity to combine and simplify these:

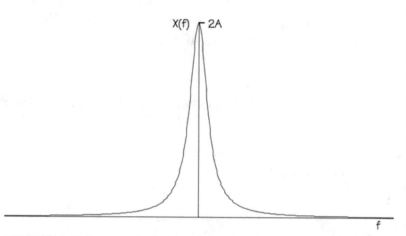

FIGURE 12.19: Spectrum for double-sided exponential.

$$X(f) = \frac{A}{a+j2\pi f} + \frac{A}{a-j2\pi f}$$

$$= \frac{2aA}{a^2+(2\pi f)^2}.$$

Fig. 12.19 shows the form of this Fourier transform. It looks very similar to the function we started with.

12.3.7 Cosine and Sine

I think it goes without saying that we need the Fourier transforms for $\cos(t)$ and $\sin(t)$. Let's start with

$$x(t) = A\cos 2\pi f_1 t.$$

If we try the integration, we'll find it's a bit of a mess. More-over, we can use our heads and get the result more simply.

The cosine has only one frequency, f_1. So its spectrum must have energy at only two points, $f = f_1$ and $f = -f_1$. (We know that spectra are symmetric.) So let's guess that this energy is in the form of a pair of impulses at $f = \pm f_1$. We'll have to have a size, so I'll guess a size of B. (See Fig. 12.20.)

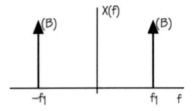

FIGURE 12.20: Cosine spectrum?

The mathematical description of this spectrum is

$$X(f) = B\delta(f + f_1) + B\delta(f - f_1).$$

Now use translation to get back to the time domain:

$$\wedge^{-1}\left[X(f - f_a)\right] = e^{+j2\pi f_a t}x(t),$$
$$\text{so } \wedge^{-1}\left[B\delta(f + f_1)\right] = Be^{-j2\pi f_1 t}\wedge^{-1}\left[\delta(f_1)\right]$$
$$= Be^{-j2\pi f_1 t}.$$

Combining the results for both impulses yields

$$x(t) = Be^{-j2\pi f_1 t} + Be^{+j2\pi f_1 t}$$
$$= 2\left(\frac{Be^{-j2\pi f_1 t} + Be^{+j2\pi f_1 t}}{2}\right)$$
$$= 2B\cos 2\pi f_1 t.$$

But we were looking for a cosine of size A, so that means that $B = A/2$. The transform of the cosine is therefore a pair of impulses:

$$\mathscr{F}\left[A\cos 2\pi f_1 t\right] = \frac{A}{2}\delta(f + f_1) + \frac{A}{2}\delta(f - f_1).$$

We can get the transform for the sine function by looking back at what we have just done. If we take the expression for $x(t)$ and modify it slightly, we can get the sine form of the expression:

$$x_s(t) = 2\left(\frac{-Be^{-j2\pi f_1 t} + Be^{+j2\pi f_1 t}}{2j}\right)$$
$$= 2B\sin 2\pi f_1 t.$$

All I did was put a minus sign in front of the first term and a j in the denominator. I'll modify the sum of the two impulses to reflect this:

$$X_s(f) = \frac{-B}{j}\delta(f + f_1) + \frac{B}{j}\delta(f - f_1)$$
$$= jB\delta(f + f_1) - jB\delta(f - f_1),$$
$$\text{so } \mathcal{F}[A\sin 2\pi f_1 t] = j\left[\frac{A}{2}\delta(f + f_1) - \frac{A}{2}\delta(f - f_1)\right].$$

The transform of the sine function is also a pair of impulses, but with a phase difference.

12.3.8 Pulse

We've already done a square pulse, but let's do it here to make the story a little more complete. The pulse is shown in Fig. 12.21. I'll make use of symmetry (this pulse has even symmetry) to simplify the integration:

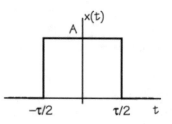

FIGURE 12.21: Pulse.

$$X(f) = \int_{-\tau/2}^{\tau/2} Ae^{-j2\pi ft}\, dt$$
$$= \int_{-\tau/2}^{\tau/2} A\left(\cos 2\pi ft + j\sin 2\pi ft\right)dt.$$

The j term in the integral is an odd function and hence will integrate to zero. So I'll throw it out, change the limits, and double the result:

$$X(f) = 2\int_{0}^{\tau/2} A\cos 2\pi ft\, dt$$
$$= \frac{2A}{2\pi f}\sin 2\pi ft\Big|_{0}^{\tau/2}$$
$$= \frac{A\tau}{\pi f \tau}\sin \pi f \tau$$
$$= A\tau\frac{\sin \pi\tau f}{\pi\tau f}.$$

The result is the familiar $(\sin x)/x$ shown in Fig. 12.22.

12.3.9 Modulation

One more form is in our "important" list—modulation. It's the heart of just about every communications system we encounter. We'll consider here just amplitude modulation, or AM. In this form of modulation, the *amplitude* of a sine wave is a function of time:

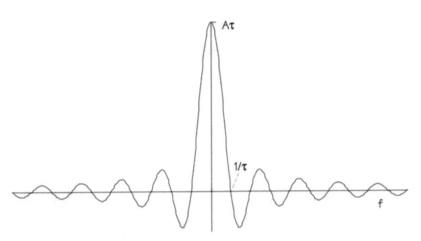

FIGURE 12.22: Spectrum for pulse A by τ.

$$x(t) = g(t)\cos 2\pi f_c t,$$

where $g(t)$ is the modulating signal and f_c is the frequency of the *carrier* that is being modulated.

I'll write this using Euler's formula:

$$\cos 2\pi f_1 t = \frac{1}{2}e^{+j2\pi f_1 t} + \frac{1}{2}e^{-j2\pi f_1 t},$$

$$x(t) = \frac{1}{2}e^{+j2\pi f_1 t}g(t) + \frac{1}{2}e^{-j2\pi f_1 t}g(t).$$

But each of these terms looks just like the result of translation in the frequency domain:

$$\mathcal{F}^{-1}\big[X[(f - f_a)]\big] = e^{j2\pi f_a t}x(t).$$

Hence the transform of amplitude modulation can be written as

$$\mathcal{F}\big[g(t)\cos 2\pi f_c t\big] = \frac{1}{2}G(f - f_c) + \frac{1}{2}G(f + f_c),$$

where $G(f)$ is the Fourier transform of $g(t)$.

This says that the spectrum of the modulating signal $g(t)$ is split into two equal parts, each half the size of the original. These are centered on the carrier frequency at $f = f_c$ and $f = -f_c$.

12.4   CIRCUITS AND FOURIER

The Fourier transform can be used to "solve" circuits problems in about the same way as we used the Laplace transform. In some ways, this isn't much different and not too useful. But in another way, the method gets some interesting results.

12.4.1 Circuit Elements

Consider the three passive elements shown in Fig. 12.23. If we presume no DC terms to confuse things, we can use the properties of the Fourier transform to get the V–I relationships. For the resistor,

$$v(t) = Ri(t),$$
$$V(f) = RI(f);$$

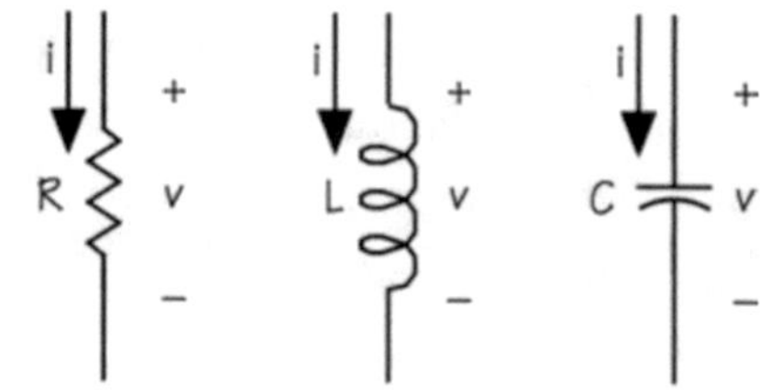

for the inductor,

$$v(t) = L\frac{di(t)}{dt},$$
$$V(f) = j2\pi fLI(f);$$

and for the capacitor,

$$i(t) = C\frac{dv(t)}{dt},$$
$$I(f) = j2\pi fCV(f).$$

We can define impedance as before:

$$Z_R = R, \quad Z_L = j2\pi fL, \quad Z_C = 1/j2\pi fC.$$

FIGURE 12.23: Circuit elements.

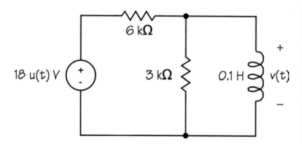

FIGURE 12.24: Circuit in the time domain.

Sources are simply transformed from the time domain using what we know about the properties of the Fourier transform and some of the basic transform pairs.

12.4.2 Example Circuit

Fig. 12.24 shows a circuit in the time domain. We are to find $v(t)$. I'll convert the circuit using the impedances we've just worked out. The source is a step and gets its funny-looking Fourier representation. Fig. 12.25 shows that conversion.

This problem would be easier to work if I converted it to its Thévenin equivalent. This conversion doesn't change in the frequency domain. The Thévenin equivalent is shown in Fig. 12.26.

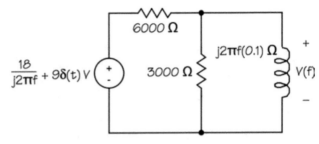

FIGURE 12.25: Circuit in Fourier terms.

Analysis is the same as in any other domain. Here we have a voltage divider, so I'll just write the voltage–divider relationship and neaten the terms:

$$V(f) = \left[\frac{6}{j2\pi f} + 3\delta(f)\right] \frac{j2\pi f(0.1)}{j2\pi f(0.1) + 2000}$$

$$= \frac{6(0.1)}{2000 + j0.2\pi f} + 3\delta(f)\frac{j0.2\pi f}{2000 + j0.2\pi f}.$$

The second term is zero everywhere except at $f = 0$, and the fractional part goes to zero there. So I'll ignore it. That leaves

$$V(f) = \frac{0.6}{2000 + j0.2\pi f},$$

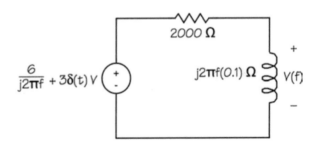

FIGURE 12.26: Thévenin in Fourier terms.

which I need to clean up to look like the transform of something that I know, namely,

$$\mathcal{F}\left[Ae^{-at}u(t)\right] = \frac{A}{a + j2\pi f}.$$

Cleaned up, the output becomes

$$V(f) = \frac{6}{20,000 + j2\pi f},$$

which transforms back to the time domain as

$$v(t) = 6e^{-20,000t}u(t)\,V.$$

Does this make sense? Let's check several points:

- At $t = 0^+$ no current can be flowing through the inductor, so the circuit is simply a voltage divider and the output should be 6 V. $v(0^+)$ is just that.
- The time constant should be L divided by the resistance the inductor "sees." That's L divided by 6 kΩ in parallel with 3 kΩ, or $0.1/2000 = 1/20,000$ s. That looks fine.
- Finally, as t goes to infinity, the current in L should be steady and the voltage across it should be zero. That checks, too.

12.4.3 Laplace in Hiding?

Is Laplace hiding in here? If we converted the circuit using what we learned in studying the Laplace transform, the circuit shown in Fig. 12.27 would be the result. Converting that to the Thévenin equivalent would be the circuit in Fig. 12.28.

Using Laplace transforms, I get the following result:

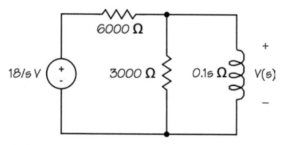

FIGURE 12.27: Circuit in s domain.

$$V(s) = \frac{6}{s}\frac{0.1s}{0.1s + 2000} = \frac{6}{s + 20000},$$

$$v(t) = 6e^{-20,000t}u(t)\,V.$$

They sure look the same!

How similar are the Fourier and the Laplace transforms? Recall that the Laplace transform was for functions that started at $t = 0$:

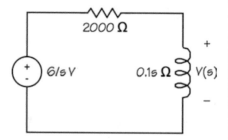

FIGURE 12.28: Thévenin equivalent.

$$X(s) = \int_{0}^{\infty} x(t)e^{-st}\,dt.$$

If a function starts at $t = 0$, the Fourier transform becomes

$$X(f) = \int_{0}^{\infty} x(t)e^{-j2\pi ft}\,dt.$$

So if $s = \sigma + j2\pi f$ (using $2\pi f$ instead of ω) and $\sigma = 0$, the two transforms will be the same. Hence, if $x(t)$ starts at $t = 0$,

$$Fourier\left[x(t)\right] = Laplace\left[x(t)\right]_{s=0}.$$

We can go the other way, too:

$$x(t) = Laplace^{-1}\left[X(j2\pi f)\big|_{j2\pi f \to s}\right]$$
$$= Fourier^{-1}\left[X(j2\pi f)\right],$$

provided the poles of the Laplace form are in the left half plane.

All of this means that Laplace transform tables can sometimes be useful in getting Fourier transforms. While this sounds good, restrictions on the Fourier transform make it some-what useless for solving transient problems. But the relationship between the transforms is an important one.

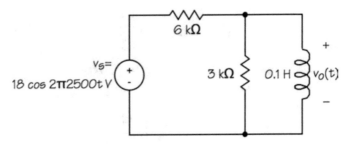

FIGURE 12.29: Circuit for the sinusoidal steady state.

12.4.4 Sinusoidal Steady State

Consider the circuit of Fig. 12.29 in the time domain. We want the output voltage $v_o(t)$ in the sinusoidal steady state.

I will transform this circuit using Fourier so that I can get the transfer function $H(f)$. The transformed circuit is shown in Fig. 12.30. The transfer function is

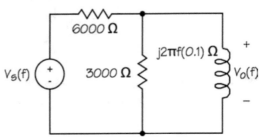

FIGURE 12.30: Circuit in the f domain.

$$H(f) = \frac{j0.2\pi f}{j0.2\pi f + 2000}.$$

Let's carry out the whole analysis:

$$V_o(f) = V_s(f)H(f),$$

$$V_s(f) = \mathcal{F}\left[\delta(f-2500) + \delta(f+2500)\right],$$

$$V_o(f) = \mathcal{F}\left[\delta(f-2500) + \delta(f+2500)\right]\frac{j0.2\pi f}{j0.2\pi f + 2000},$$

$$v_o(t) = \mathcal{F}^{-1}\left[V_o(f)\right]$$

$$= \int_{-\infty}^{\infty} \mathcal{F}\left[\delta(f-2500) + \delta(f+2500)\right]\frac{j0.2\pi f}{j0.2\pi f + 2000}e^{+j2\pi ft}\,df.$$

Now recall that the impulse lets us "see through it" only at the point where it is active. Hence the terms inside the integral are evaluated at $f = +2500$ and $f = -2500$. Now I can finish the job:

$$v_o(t) = \mathcal{F}\left[\frac{j0.2\pi 2500}{j0.2\pi 2500 + 2000}e^{+j2\pi 2500t}\right.$$

$$\left. + \frac{j0.2\pi(-2500)}{j0.2\pi(-2500) + 2000}e^{-j2\pi 2500t}\right]$$

$$= \left[\frac{1571\angle 90°}{2543\angle 38.1°}e^{+j2\pi 2500t}\right.$$

$$\left. + \frac{1571\angle -90°}{2543\angle -38.1°}e^{-j2\pi 2500t}\right]$$

$$= 5.56e^{+j2\pi 2500t}\angle 51.9° + 5.56e^{-j2\pi 2500t}\angle -51.9°$$

$$= 11.12\cos(2\pi 2500t + 51.9°)\,\text{V}.$$

Wow! We got the result for the sinusoidal steady state directly from the transformed circuit. But was it easier than what we have been doing with phasors? I'll let you decide!

12.5   ENERGY

We already know from our previous work that the spectrum carries energy information. We even simplified this by standardizing on the "1-Ω energy," applying the voltage to a 1-Ω resistor and computing the power.

But do Fourier transforms and the spectra we plot from these carry this energy information? It seems obvious that this should be true, but it might not be a bad idea to prove it.

Our spectra are usually *density* spectra. We plot such quantities as *volts per hertz*. When we plot the 1-Ω energy, we plot *volts squared per hertz*. When we integrate such a spectrum over a frequency range, we get *joules per hertz*. Now let's see where the mathematics takes us.

Parseval's theorem (no connection to any of Herr Wagner's overlong operas) says that we can get the 1-Ω energy directly from the Fourier transform. The integration is

$$W_{1\Omega} = \int_{\substack{desired \\ frequency \\ range}} |X(f)|^2 \, df.$$

That's certainly no surprise! We already knew this from stuff we have done before. Or did we? Are we taking this on faith. A simple proof is in order, starting with the 1-Ω energy in the time domain:

$$W_{1\Omega} = \int_{-\infty}^{\infty} x^2(t) dt$$

$$= \int_{-\infty}^{\infty} x(t)x(t) dt.$$

Now replace the second appearance of $x(t)$ with its inverse Fourier transform and rearrange the order of integration:

$$W_{1\Omega} = \int_{-\infty}^{\infty} x(t) \left[\int_{-\infty}^{\infty} X(f)e^{+j2\pi ft} \, df \right] dt$$

$$= \int_{-\infty}^{\infty} \int_{-\infty}^{\infty} x(t)X(f)e^{+j2\pi ft} \, df \, dt$$

$$= \int_{-\infty}^{\infty} \int_{-\infty}^{\infty} X(f)x(t)e^{+j2\pi ft} \, dt \, df$$

$$= \int_{-\infty}^{\infty} X(f) \left[\int_{-\infty}^{\infty} x(t)e^{-j2\pi(-f)t} \, dt \right] df.$$

The integration in the brackets is the Fourier transform of $x(t)$, but with f replaced by $-f$. If we use this fact we can finish the job. We also need the fact that the product of $X(f)$ and $X(-f)$ is the magnitude squared of $X(f)$, something we saw when we looked at the even and odd properties of the Fourier transform.

$$W_{1\Omega} = \int_{-\infty}^{\infty} X(f)X(-f) df$$

$$= \int_{-\infty}^{\infty} |X(f)|^2 \, df,$$

which is what we set out to prove.

To summarize Parseval's theorem,

$$W_{1\Omega} = \int_{-\infty}^{\infty} x^2(t)\,dt = \int_{-\infty}^{\infty} |X(f)|^2\,df.$$

Since this function in the frequency domain is an even function, we can cut the range of integration in two:

$$W_{1\Omega} = 2\int_{0}^{\infty} |X(f)|^2\,df.$$

We can also find such results as the 1-Ω energy in a certain frequency band:

$$W_{1\Omega} = 2\int_{f_1}^{f_2} |X(f)|^2\,df.$$

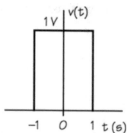

FIGURE 12.31: Two-second pulse.

Let's do an example using the pulse of Section 12.2, repeated here in Fig. 12.31. The pulse is

$$v(t) = \begin{cases} 1 & -1 < t < 1\,\mathrm{s} \\ 0 & \text{elsewhere} \end{cases}.$$

How much energy is contained in the pulse? This is easier to do in the time domain:

$$W_{1\Omega} = 2\int_{0}^{1} 1^2\,dt = 2t\big|_{0}^{1} = 2\,\mathrm{J}.$$

What percentage of this total energy is contained in the "first lobe" of the sinx/x spectrum (Fig. 12.32)? The first lobe is the large bump, so we need to see where the first zeros are.

Look at the frequency-domain function. Note that the first place where this can go to zero is when the sin has a zero. That will be when its argument equals π:

$$V(f) = 2\frac{\sin 2\pi f}{2\pi f},$$
$$\sin 2\pi f = 0,$$
$$2\pi f = \pi,$$
$$f = 0.5\,\mathrm{s}.$$

I'll use Parseval's theorem and integrate over the first lobe:

$$W_{1\Omega} = 2\int_0^{0.5} \left| 2\frac{\sin 2\pi f}{2\pi f} \right|^2 df$$

$$= 1.806 \text{ J.}$$

$$percent = 100\frac{1.806}{2} = 90.3\%.$$

This says that if we were to pass this pulse through an ideal low-pass filter whose cutoff frequency is 0.5 Hz, 90% of the energy of the pulse would pass through. (For

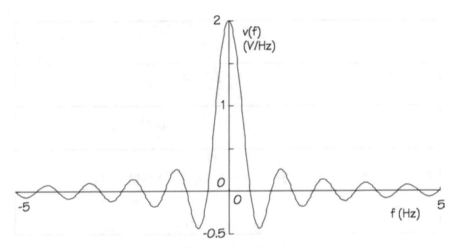

FIGURE 12.32: Fourier transform of two-second pulse.

the record, if we include the second lobe by making the cutoff frequency 1 Hz, we would get 95% of the original energy.)

Parseval said something we have already known, namely, that we can get the 1-Ω energy of a signal from its spectrum. All we need to do is integrate the square of the magnitude of the transform over the desired frequency range.

12.6 FOURIER AND CONVOLUTION

When I was a student encountering convolution for the first time, only three things stuck in my mind: 1) I don't understand this, 2) I don't see what it's good for, and 3) I never want to see this again! But if you'll have faith and stick with me for a while, I think you'll see what the rudiments of convolution are and even what it is good for. (I can't do anything about the third point, though.)

I'll start by asking two questions. First, if

$$x(t) = x_1(t)x_2(t),$$

what is its Fourier transform $X(f)$? Second, if

$$X(f) = X_1(f)X_2(f),$$

what is $x(t)$ in the time domain? The answers to both questions turn out to be useful. (Amplitude modulation involves multiplication in the time domain, for example.) Needless to say, both answers involve convolution.

Let's start this in the time domain. Suppose

$$x_1(t) = \int_{-\infty}^{\infty} X_1(f)e^{+j2\pi ft}\, df.$$

I'll substitute the dummy variable ϕ for f in this integration:

$$x_1(t) = \int_{-\infty}^{\infty} X_1(\phi)e^{+j2\pi\phi t}\, d\phi.$$

Now I multiply both sides by $x_2(t)$ and group this with the exponential:

$$x_1(t)x_2(t) = \int_{-\infty}^{\infty} X_1(\phi)\left[e^{+j2\pi\phi t}x_2(t)\right]d\phi.$$

That term in the bracket is translation in the frequency domain. In other words, it is the inverse transform of $X_2(f)$ shifted on the f axis by a distance of ϕ. I can combine all this into one integral:

$$e^{+j2\pi\phi t}x_2(t) = \wedge^{-1}\left[X_2(f-\phi)\right]$$

$$= \int_{-\infty}^{\infty} X_2(f-\phi)e^{+j2\pi ft}\, df,$$

$$x_1(t)x_2(t) = \int_{-\infty}^{\infty} X_1(\phi)\left[\int_{-\infty}^{\infty} X_2(f-\phi)e^{+j2\pi ft}\, df\right]d\phi$$

Now reorder the terms of the integrations to get f on the outside and ϕ on the inside:

$$x_1(t)x_2(t) = \int_{-\infty}^{\infty}\left[\int_{-\infty}^{\infty} X_1(\phi)X_2(f-\phi)d\phi\right]e^{+j2\pi ft}\, df.$$

The outer integral creates the inverse Fourier transform of the term in the brackets. The stuff inside that bracket is *convolution*, even though we might be hardpressed to discover that at the moment. So I'll just say, take it on faith that the name of the stuff in the brackets is convolution! (The asterisk is used as the mathematical symbol for convolution, not to be confused with multiplication in computer languages or with complex conjugates.)

$$\mathcal{F}\left[x_1(t)x_2(t)\right] = \int_{-\infty}^{\infty} X_1(\phi)X_2(f - \phi)d\phi$$

$$= X_1(f) * X_2(f).$$

If we played the same game but start in the frequency domain, we get convolution in the other direction:

$$\mathcal{F}^{-1}\left[X_1(f)X_2(f)\right] = \int_{-\infty}^{\infty} x_1(\alpha)x_2(t - \alpha)d\alpha$$

$$= x_1(t) * x_2(t).$$

Are you still alive? OK, then why don't you ask what good this is?

Thank you for asking! Convolution with impulses is most useful. Uhuh! So? Well, let's see what happens by carrying out convolution (yes, actually doing the integration) with a square pulse. That'll do two things. First, we'll see how one carries out the process. Second, by letting the pulse width go to zero while keeping the same area, we'll show the result of convolution with an impulse.

Fig. 12.33 shows two functions. I wish to convolve them:

$$y(t) = x_1(t) * x_2(t)$$

$$= \int_{-\infty}^{\infty} x_1(\alpha)x_2(t - \alpha)d\alpha.$$

I now modify these functions to fit what is needed to carry out the convolution integral:

1. Rewrite both functions as functions of α instead of t (Fig. 12.34).

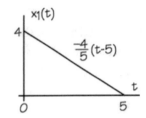
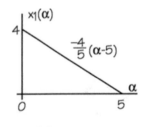

FIGURE 12.33: Two functions.

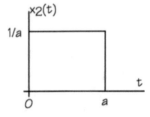
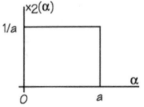

FIGURE 12.34: Functions in α.

2. Now flip x_2 so that it is a function of $-\alpha$ instead of $+\alpha$ (Fig. 12.35).

3. Finally, add the variable t to the argument, thereby shifting the pulse so that its right edge is at t rather than at $\alpha = 0$ (Fig. 12.36).

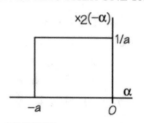

FIGURE 12.35: $x_2(\alpha)$ flipped.

Now I am ready to carry out the integration. I must integrate over all values of α while sliding the flipped function $x_2(t-\alpha)$ from left to right along the same axis as $x_1(\alpha)$. In other words, I'll put both functions on the same α axis and move x_2 relative to x_1. The position of x_2 is governed by the value of t; the position of x_1 is fixed in its original position.

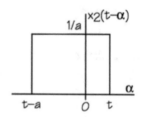

FIGURE 12.36: $x_2(-\alpha)$ shifted.

There are five separate cases to consider: 1) no overlap, 2) partial overlap from the left, 3) pulse completely inside the triangle, 4) partial overlap at the right, and 5) no overlap again.

Case 1: Here $t < 0$ so the pulse is sitting off to the left of the triangle. There is no overlap, so the product of the two is zero (Fig. 12.37):

$$y(t) = 0.$$

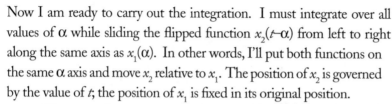

FIGURE 12.37: Convolution Case 1.

Case 2: The pulse is starting to overlap the left side of the triangle but it isn't fully inside yet. This is for $0 < t < a$. The product is integrated from $\alpha = 0$ to $\alpha = t$ (Fig. 12.38):

$$y(t) = \int_0^t \frac{1}{a}\left[-\frac{4}{5}(\alpha - 5)\right] d\alpha$$

$$= -\frac{2}{5}\frac{t(t-10)}{a}.$$

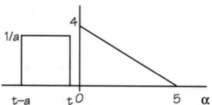

FIGURE 12.38: Convolution Case 2.

Case 3: The pulse is entirely inside the triangle, so this means $a < t < 5$. There is a product and I integrate between $t-a$ and t (Fig. 12.39):

$$y(t) = \int_{t-a}^t \frac{1}{a}\left[-\frac{4}{5}(\alpha - 5)\right] d\alpha$$

$$= 4 - \frac{4}{5}t + \frac{2}{5}a.$$

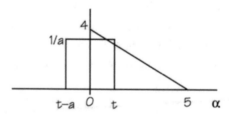

FIGURE 12.39: Convolution Case 3.

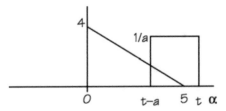

FIGURE 12.40: Convolution Case 4.

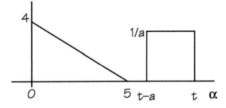

FIGURE 12.41: Convolution Case 5.

Case 4: The pulse is beginning to escape on the right. The range of t is now $5 < t < 5+a$. There is still a product but only between $\alpha = t-a$ and $\alpha = 5$ (Fig. 12.40):

$$y(t) = \int_{t-a}^{5} \frac{1}{a}\left[-\frac{4}{5}(\alpha - 5)\right] d\alpha$$

$$= \frac{2}{5}\frac{25 - 10t + t^2 - 2ta + 10a + a^2}{a}.$$

Case 5: The pulse has completely escaped so that there is again no overlap and no product. Hence $5+a < t$ and the integral is zero (Fig. 12.41):

$$y(t) = 0.$$

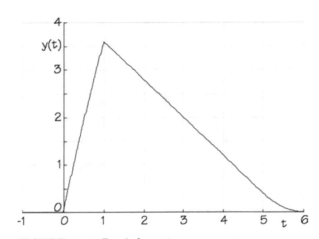

FIGURE 12.42: Result for $a = 1$.

I'm not going to collect all of those pieces into one statement, but I am going to plot the result. The graph of Fig. 12.42 is plotted with $a = 1$. Does that look like the original triangle? No, not really.

Now look at the plot of Fig. 12.43, where $a = 0.1$. Closer to the triangle, right? And how about Fig. 12.44 for $a = 0.01$? It's pretty hard to tell this from the original triangle.

So we get to the remarkable conclusion that convolution with an impulse at $t = 0$ gives us the original function

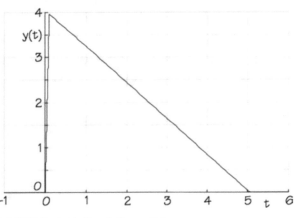

FIGURE 12.43: Result for $a = 0.1$.

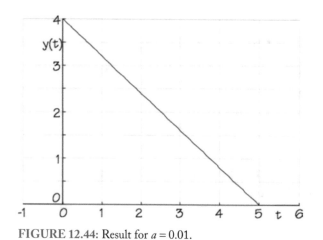

FIGURE 12.44: Result for $a = 0.01$.

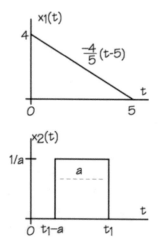

FIGURE 12.45: Two functions again.

back. Oh, no, really? If you are thinking that we could have just kept the original function and not gone to all this trouble....

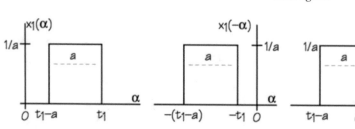

FIGURE 12.46: $x_1(t)$ "alpha'd," flipped, and shifted by t.

But that was so much fun that I want to do it all over again. This time, though, I'll plant the right edge of the pulse at the time t_1 as shown in Fig. 12.45. Then I'll convert things to α, flip x_2, and slide it along the α axis by an amount t. (See Fig. 12.46.)

Note that the leading (right) edge of the pulse is now at $t - t_1 + a$. My work proceeds in five steps as before:

Case 1: No overlap, $t - t_1 + a < 0$ or in terms of t, $t < t_1 - a$ (Fig. 12.47):

$$y(t) = 0.$$

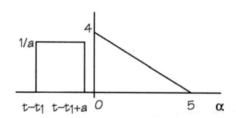

FIGURE 12.47: Convolution Case 1 again.

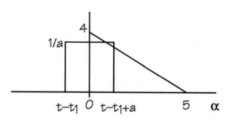

FIGURE 12.48: Convolution Case 2 again.

Case 2: Partial overlap on the left, so $0 < t - t_1 + a < t - t_1$ or in terms of t, $t_1 - a < t < t - a$ (Fig. 12.48):

$$y(t) = \int_0^{t-t_1+a} \frac{1}{a}\left[-\frac{4}{5}(\alpha - 5)\right]d\alpha.$$

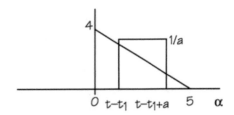

FIGURE 12.49: Convolution Case 3 again.

Case 3: Complete overlap where $a < t - t_1 + a < 5$ or in terms of t, $t_1 < t < 5 + t_1 - a$ (Fig. 12.49):

$$y(t) = \int_{t-t_1}^{t-t_1+a} \frac{1}{a}\left[-\frac{4}{5}(\alpha - 5)\right]d\alpha.$$

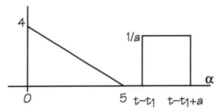

FIGURE 12.50: Convolution Case 4 again.

Case 4: Partial overlap on the right where $5 < t - t_1 + a < 5 + t - t_1$ or in terms of t, $5 + t_1 - a < t < 5 + t_1$ (Fig. 12.50):

$$y(t) = \int_{t-t_1}^5 \frac{1}{a}\left[-\frac{4}{5}(\alpha - 5)\right]d\alpha.$$

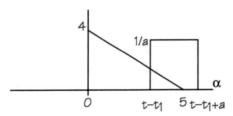

FIGURE 12.51: Convolution Case 5 again.

Case 5: No overlap again, so $5 + t - t_1 < t - t_1 + a$ or in terms of t, $5 + t_1 < t$ (Fig. 12.51):

$$y(t) = 0.$$

I'll save a few drawings by plotting the results of all this work with $a = 0.01$. In Fig. 12.52 I have set $t_1 = 10$ to give the pulse a fixed position. Note that this is the original triangle, but instead of beginning at $t = 0$, it begins at $t = 10$, the value I set for t_1.

My conclusion? An important one! Convolution of a function with an impulse at $t = t_1$ recreates the original function, but now it starts t_1 units of time later. In "math" terms, this is

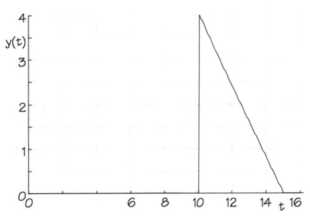

FIGURE 12.52: Result for $a = 0.01$ and $t_1 = 10$.

$$y(t) = \delta(t - t_1) * x(t)$$
$$= x(t - t_1).$$

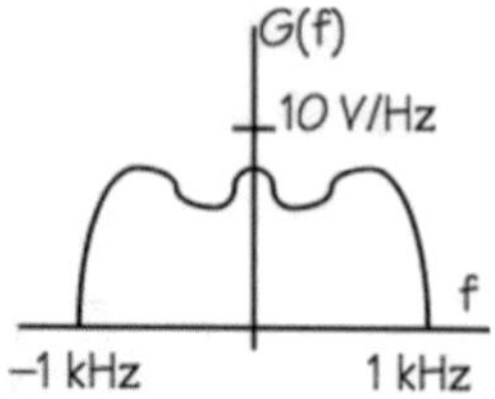

Here's one more example, this one more closely related to what we have been doing. Consider amplitude modulation again:

$$x(t) = g(t)\cos 2\pi f_c t,$$

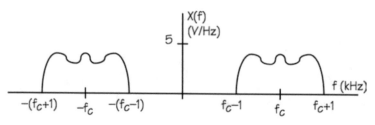

FIGURE 12.53: Base-band spectrum.

where $g(t)$ is the modulating function (i.e., the amplitude of the cosine) and f_c is the frequency of the carrier (the cosine).

Suppose that we know the spectrum of the modulation. In other words, suppose we know $G(f)$ for $g(t)$, something that we often *do* know. Figure 12.53 shows a possible spectrum.

The spectrum for the cosine is easy, just a couple of impulses at $-f_c$ and $+f_c$ as shown in Fig. 12.54.

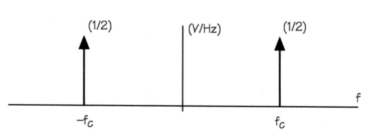

FIGURE 12.54: Spectrum of cosine.

FIGURE 12.55: Result of convolution.

Our modulation operation in the time domain is multiplication. This becomes convolution in the frequency domain:

$$X(f) = G(f) *_\wedge \left[\cos 2\pi f_c t\right].$$

But this is just convolution with two impulses. Linearity says we can do each one separately and add them together. Convolution with an impulse returns the original function but shifted by the position of the impulse. So the result, shown in Fig. 12.55, is the original spectrum of $g(t)$ repositioned on the frequency axis in two places, $-f_c$ and $+f_c$.

That's a lot easier than trying to carry out the mathematics of multiplying the two functions in the time domain and then trying to compute their Fourier transform! (If you don't believe this, take a simple function for $g(t)$ such as a sine and try it the long way!)

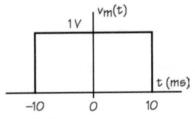

FIGURE 12.56: Code pulse.

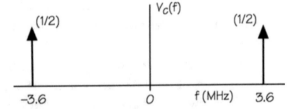

FIGURE 12.57: Spectrum of carrier.

12.7   DESIGN EXAMPLE

A ham is having a QSO with another ham on CW near the bottom of the 80-m band. Translation: Two amateur radio operators are communicating in Morse code with a carrier frequency of 3.6 MHz. One ham is sending one dot, the code for the letter e. Let's see first what her signal looks like.

Fig. 12.56 shows the pulse for this dot. It has an amplitude of 1 V and a width of 20 ms, centered at $t = 0$. Fig. 12.57 shows spectrum of the transmitter's unmodulated carrier.

We already know how to find the spectrum for her signal by using convolution with impulses. Her dot pulse has the spectrum shown in Fig. 12.58. Convolving this with the spectrum of Fig. 12.57 gives the result shown in Fig. 12.59. (I have cheated and distorted the frequency axis so we can see what is going on.)

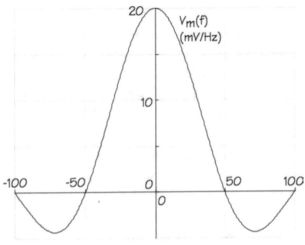

FIGURE 12.58: Spectrum of the code pulse.

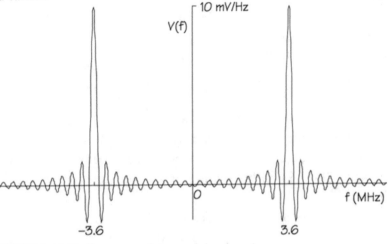

FIGURE 12.59: Spectrum of the modulated carrier.

Our ham operator decides that she'd like to have the bandwidth of her signal reduced. One reason is that a narrower signal can often get through on a band where there are lots of other signals and noise.

That means we need to analyze her signal. I'll choose to look for the frequency range over which 90% of the 1-Ω energy is concentrated. But the beauty of what we have been learning is that we don't have to do this with the complicated spectrum of Fig. 12.59! Convolution with an impulse just returns the original function, perhaps shifted. So we can do all our studies using the baseband signal spectrum of Fig. 12.58. That makes life a lot simpler.

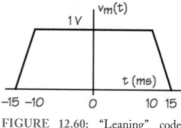

FIGURE 12.60: "Leaning" code pulse.

Recall that we did an example in Section 12.5 that showed that about 90% of the signal's 1-Ω energy was concentrated in the first lobe of the $(\sin x)/x$ spectrum. This says that our ham can filter her signal to limit it to frequencies below 50 Hz and still have most of the energy.

Now suppose she decides, since she now knows something about Fourier and spectra, that she could improve things by smoothing out the dot waveform a bit. After all, the sharper the corners, the wider the spectrum.

This leads her to design a circuit that slopes the sides of the pulse as shown in Fig. 12.60. The Fourier transform gives the spectrum shown in Fig. 12.61.

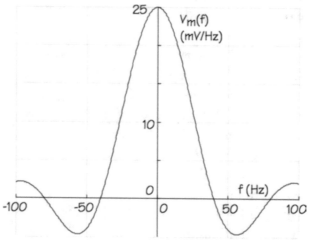

FIGURE 12.61: Spectrum of the "leaning" code pulse.

She integrates the square of the magnitude of the spectrum between various limits until she finds that the 90% bandwidth is between −25 and + 25 Hz. Gosh, that is a much smaller bandwidth.

That came out so good that she decides to "lean" the edges all the way and make the dot into the triangle shown in Fig. 12.62. The Fourier spectrum for this one is given in Fig. 12.63.

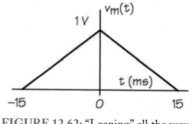

FIGURE 12.62: "Leaning" all the way.

Again she finds the bandwidth of 90% of the 1-Ω energy by several trial-and-error integrations. The result is 28 Hz. No improvement—even worse, in fact.

Several points are worth making:

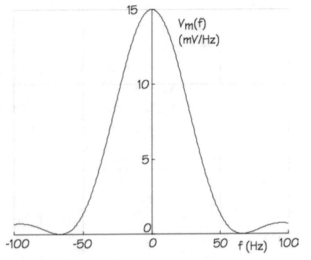

FIGURE 12.63: Spectrum of the "all-the-way" pulse.

- Smoothing a signal narrows the bandwidth.
- Working with the baseband signal and its transform is easier than doing it with the modulated signal.
- The 90%-energy bandwidth can generally be found by trial-and-error integration of the square of the magnitude of the spectrum.

12.8 SUMMARY

This chapter turned out to be a mixture of signals and math and some other strange stuff that may not seem all that related. Yet this is all built around the Fourier transform.

When we had periodic signals, we could use the Fourier series to transform these signals into the frequency domain. The result was a discrete spectrum, a spectrum that had "values" only at specific frequencies. And these "specific frequencies" were harmonics of the frequency of the original signal.

If we don't have periodic signals, the Fourier transform gives us a continuous spectrum. Now we can transform an individual pulse into a spectrum. These spectra often seem to involve $(\sin x)/x$, which we note has all frequencies present, all the way to infinity.

We should not have been surprised to learn that a signal that has abrupt changes has a wide frequency spectrum. After all, if we want to change something infinitely fast, we ought to expect lots of "fast" energy, energy that appears "way out" in the spectrum.

Our goal in this chapter has been to introduce the Fourier transform and continuous spectra without all sorts of mathematics. You might say that I didn't succeed. But my goal was to add tools to your tool box so you could find spectra *without* lots of math. To this end, we worked through a number of properties with the idea that these could help us avoid much of the math. Table 12.1 summarizes most of what we've developed.

Now that you can go between the time domain and the frequency domain in all sorts of situations...gosh, we can do this with circuits, with filters, using s, using f, using $j2\pi f$, through sketches of Bode diagrams, with periodic signals, with non-periodic signals, with....

Don't be surprised when this stuff keeps resurfacing. Yes, I know, the world is heavily "into digital." But those pulses still have to move through physical channels made out of physical elements. Even digitial signal processing involves heavy computation using algorithms such as the fast Fourier transform. Analog circuits and analog signals aren't going away any time soon!

TABLE 12.1: Fourier transforms

Fourier transform	$X(f) = \int_{-\infty}^{\infty} x(t)e^{-j2\pi ft}\,dt$
Inverse transform	$x(t) = \int_{-\infty}^{\infty} X(f)e^{+j2\pi ft}\,dt$
For even function	$A(f) = \int_{-\infty}^{\infty} x(t)\cos(2\pi ft)\,dt$
For odd function	$B(f) = -\int_{-\infty}^{\infty} x(t)\sin(2\pi ft)\,dt$
Linearity	$ax_1(t) + bx_2(t) \leftrightarrow aX_1(f) + bX_2(f)$
Scaling (numbers on t axis multiplied by a, so function is "slower")	$\mathcal{F}[x(at)] = \dfrac{1}{a}X\left(\dfrac{f}{a}\right)$
Scaling (numbers on f axis multiplied by a, so frequencies are higher)	$X(af) = a\mathcal{F}\left[x\left(\dfrac{t}{a}\right)\right]$
Flipping (t or f flipped end for end)	$\mathcal{F}[x(-t)] = X(-f)$
Shifting (starting point delayed by a)	$\mathcal{F}[x(t-a)] = e^{-j2\pi fa}\mathcal{F}[x(t)]$
Shifting (moved f_a to right on f axis)	$\mathcal{F}^{-1}[X(f-f_a)] = e^{j2\pi f_a t}x(t)$
Impulse (in either domain)	$\begin{cases} \mathcal{F}[A\delta(t-t_a)] = Ae^{-j2\pi ft_a}, \\ \mathcal{F}^{-1}[B\delta(f-f_b)] = B\cos 2\pi f_b t + jB\sin 2\pi f_b t \end{cases}$
Constant (in either domain)	$\mathcal{F}[A] = A\delta(f), \ \mathcal{F}^{-1}[A] = A\delta(t)$

Signum	$\mathcal{F}[A\, signum(t)] = \dfrac{2A}{j2\pi f}$
Differentiations	$\mathcal{F}\left[\dfrac{dx(t)}{dt}\right] = j2\pi f\,\mathcal{F}[x(t)]$
Integration	$\mathcal{F}\left[\displaystyle\int_{-\infty}^{\infty} x(\tau)d\tau\right] = \dfrac{1}{j2\pi f}\,\mathcal{F}[x(t)] + \dfrac{1}{2}x(0)\delta(f)$
Exponential (double-sided: $e^{-a|t|}$)	$X(f) = \dfrac{2aA}{a^2 + (2\pi f)^2}$
Cosine	$\mathcal{F}[A\cos 2\pi f_1 t] = \dfrac{A}{2}\delta(f+f_1) + \dfrac{A}{2}\delta(f-f_1)$
Sine	$\mathcal{F}[A\sin 2\pi f_1 t] = j\dfrac{A}{2}\delta(f+f_1) - j\dfrac{A}{2}\delta(f-f_1)$
Pulse (rectangular)	$x(t) = \begin{cases} A & -\tau/2 < t < \tau/2 \\ 0 & \text{elsewhere} \end{cases}$, $\quad X(f) = A\tau\dfrac{\sin \pi\tau f}{\pi\tau f}$
Modulation (AM)	$x(t) = g(t)\cos 2\pi f_c t,\ X(f) = \dfrac{1}{2}G(f+f_c) + \dfrac{1}{2}G(f-f_c)$
Convolution	$\mathcal{F}[x_1(t)x_2(t)] = X_1(f) * X_2(f) = \displaystyle\int_{-\infty}^{\infty} X_1(\phi)X_2(f-\phi)d\phi$ $\mathcal{F}^{-1}[X_1(f)X_2(f)] = x_1(t) * x_2(t) = \displaystyle\int_{-\infty}^{\infty} x_1(\tau)x_2(t-\tau)d\tau$

Biography

Bill Eccles has been Professor of Electrical and Computer Engineering at Rose-Hulman Institute of Technology since 1990 (except for one year at Oklahoma State). He retired in 1990 as Distinguished Professor Emeritus after 25 years at the University of South Carolina. He founded the Department of Computer Science at that university, and served at one time or another as head of four different departments, Computer Science, Mathematics and Computer Science, and Electrical and Computer Engineering, all at South Carolina, and Electrical and Computer Engineering at Rose-Hulman. Most of his teaching has been in circuits and in microprocessor systems. He has published Microprocessor Systems: A 16-Bit Approach (Addison-Wersley, 1985) and numerous monographs on circuits, systems, microprocessor programming, and digital logic design. Bill learned circuit theory at M.I.T. under Ernest Guillemin, one of the pioneers in modern circuit theory, and William Hayt at Purdue University. Bill and his wife Trish have two children and three grandchildren. Bill is also a conductor (appropriate for an electrical engineer) on the Whitewater Valley Railroad, a tourist line in Connersville, Indiana. He is a Registered Professional Engineer and an amateur radio operator.

Printed in the United States
by Baker & Taylor Publisher Services